Critical Network Infrastructures

Highlighting critical issues surrounding network infrastructures, this book offers innovative methods and parameters for modelling, analysing, enhancing reliability and safety, and addressing vulnerabilities. It is imperative to develop new tools or enhance existing ones to ensure the dependable, secure, and resilient operation of these infrastructures, and this book offers solutions to achieve this goal. The primary focus of the book lies in the application of graph theory and game theory and methods in various capacities. Drawing from over 15 years of research, the authors have distilled their findings into simple yet powerful tools for mathematical modelling, analysis, and optimisation of real, intricate technical systems and networks.

This book presents a straightforward approach to graph theory for modelling, analysing, and optimising various complex networks. It explores weight functions within graph theory and utilises graph theory terminology to assess the reliability, safety, and vulnerability of network structures. The text combines SWOT analysis and game theory with nature, graph and game theories, incorporating various multi-criteria methods and other strategies for continuous goal functions. The book continues by introducing readers to a range of system and network modelling tools, including graph and game theory models, along with algorithms, methods, and techniques for both single and multi-criteria optimisation.

Written to be used as a reference for specialists, engineers, and academics involved in research and development. It also serves professionals responsible for the reliable and safe operation of transportation systems, energy networks, and telecommunication networks, as well as those developing solutions for Intelligent Transport Networks.

Critical Network Infrastructures

Methods for Modelling and Analysis

Sambor Guze

CRC Press
Taylor & Francis Group
Boca Raton London New York

CRC Press is an imprint of the
Taylor & Francis Group, an **informa** business

Designed cover image: Shutterstock

First edition published 2026
by CRC Press
2385 NW Executive Center Drive, Suite 320, Boca Raton FL 33431

and by CRC Press
4 Park Square, Milton Park, Abingdon, Oxon, OX14 4RN

CRC Press is an imprint of Taylor & Francis Group, LLC

The publication is financed by the Dean of the Faculty of Navigation, Gdynia Maritime University and funds from the Faculty grant number: WN\PZ\06\2025

For Product Safety Concerns and Information please contact our EU representative GPSR@taylorandfrancis.com. Taylor & Francis Verlag GmbH, Kaufingerstraße 24, 80331 München, Germany.

ISBN: 978-1-041-11668-4 (hbk)
ISBN: 978-1-041-11742-1 (pbk)
ISBN: 978-1-003-66145-0 (ebk)

DOI: 10.1201/9781003661450

Typeset in CMR10 font
by KnowledgeWorks Global Ltd.

Publisher's note: This book has been prepared from camera-ready copy provided by the authors.

Contents

Preface

This book presents issues related to critical network infrastructures and provides methods for modelling, analysis, reliability, safety, and vulnerability improvement.

In everyday life, the ability to ensure the functioning of societies is essential. Civilisation's progress and globalisation have made access to transport, energy, and IT infrastructure in all countries of the world important, in addition to food security. In the era of climate change and geopolitical uncertainty, it is crucial to provide new or improved existing tools that support the reliable, safe, and resilient operation of the above mentioned infrastructures.

This book aims to provide such methods. However, we cannot forget that the simpler the tool, the greater its chance of becoming popular. That is why I placed the main methodological burden on graph theory and game theory. The stochastic, probabilistic, and statistical methods will be helpful in various forms; however, in this book, only these elements are necessary to understand the new approaches. There is no detailed description of them.

The book is divided into five chapters.

The first chapter introduces the reader to critical infrastructures and their protection. To this end, it presents these concepts in relation to four locations: the European Union, the United States of America, the United Kingdom, and Poland. It then seamlessly transitions into the concept of Critical Network Infrastructures. These comprise three basic networks: power, road, and ICT. The chapter then explores resilience, vulnerability, and risk in relation to technical systems and critical infrastructures. The final section discusses the fundamentals of reliability for complex technical systems, including two-state, ageing, multi-state, non-renewable, and renewable systems. This chapter provides a general introduction to the rest of the book.

The second chapter introduces the reader to the world of mathematical modelling of critical network infrastructures. It begins with a discussion of basic modelling principles and model categories. This clarifies the purpose of the models introduced in subsequent sections. The primary goal of this book is to familiarise and convince the broadest possible audience to apply the methods proposed here. Therefore, the second chapter focuses on graph theory and game theory models. The former were used to create a specific language for describing the reliability of all network structures using concepts derived from graph theory. This resulted in a topological description of the reliability of critical network infrastructures – an easy-to-use approach. The fundamentals of game theory, in turn, provide decision models and methods for analysing critical network infrastructures, developed in Chapter 3. The chapter concludes with three case studies. Two refer to oil pipeline networks in the USA and Saudi Arabia. The third concerns a model of an intersection in the city of Gdynia, in northern Poland.

Chapter Three describes methods for analysing critical network infrastructures based on the models introduced in Chapter Two. These primarily focus on vulnerability and resilience analysis. To this end, the concepts of structural vulnerability and weighted structural vulnerability are defined. New graph theory parameters are also determined to support the characterisation of the vulnerability and resilience of critical network infrastructures. The new parameters are: bondage connected number, connected bondage number, and weighted bondage connected number. Each of these parameters is interpreted in the previously mentioned characteristics. This chapter also introduces a new method for solving games with nature, preceded by a SWOT analysis. It is also worth emphasising that a combination of the firefighter algorithm and the two-person zero-sum attacker-defender game model is proposed. This creates a tool for simulating the spread of threats in a network, with the ability to calculate potential losses/gains from mitigation actions. This allows for the analysis of the effectiveness and cost-effectiveness of crisis scenarios. The chapter concludes with case studies of three networks: the two previously discussed in Chapter 2 and the Baltic Sea port network. These are analysed for vulnerability using the methods introduced earlier in this chapter.

Chapter four addresses the topic of optimising critical network infrastructures. After introducing basic concepts such as single-objective and multi-objective optimisation and Pareto optimality, it moves on to propose a new approach to optimising critical network infrastructures. The main goal of this chapter is to expand the range of available methods. For discrete multi-objective optimisation, a binary knapsack problem is proposed. This eliminates the need for bi-level programming. Furthermore, the new families of weighted graphs with continuous weighting functions defined in Chapter 2 open up the possibility of using linear and nonlinear multi-objective optimisation methods, as well as single-objective optimisation methods. Finally, the chapter presents examples of the application of these methods in academic settings.

Chapter five summarizes the book. It outlines what has been achieved and what remains to be done on the topic.

This book is the result of the author's more than 15 years of research on simple tools for mathematical modelling, analysis, and optimisation of real, complex technical systems and networks.

Index of acronyms, abbreviations and symbols

Most important abbreviations and acronyms

BPN – Baltic Ports Network
CEWF – continuous-edge-weighted function
CEWG – continuous-edge-weighted graph
CVWF – continuous-vertex-weighted function
CVWG – continuous-vertex-weighted graph
CER – Critical Entities Resilience
CI – Critical Infrastructure
CNI – Critical Network Infrastructure
DORA – Digital Operational Resilience Act
ECI – European Critical Infrastructure
EMV – Expected Monetary Value
EOL – Expected Opportunity Loss
EU – European Union
EVPI – Expected Value of Perfect Information
EVWPI – Expected Value of the outcome With Perfect Information
EVWOPI – Expected Value Without Perfect Information
HLGA – Hajela and Lin's Weighting-based Genetic Algorithm
LAN – Local Area Network
MAN – Metropolitan Area Network
mbv – maximal burned vertices
MDT – mean downtime
MRT – mean repair time
msv – maximal number of saved vertices
MTBF – Mean Time Between Failures
MTTF – mean time to failure
MTTR – mean time to repair
MTTV – mean time to recovery
MUT – mean uptime
NIS – network and information systems
NPGA – Niched Pareto Genetic Algorithm
SPEA – Strength Pareto Evolutionary Algorithm
SWOT – Strengths, Weaknesses, Opportunities, and Threats
WAN – Wide Area Network
VEGA – Vector Evaluated Genetic Algorithm

Most important symbols

A_i – a non-empty set of player i actions (pure strategies)
A_I – the event that all elements in the $I-th$ smallest set of paths are healthy
$\overline{A_I}$ – the complementary event of A_I
$b(G)$ – bondage number
$b_c(G)$ – bondage connected number
$cb(G)$ – connected bondage number
C_n – cyclic graph with n vertices/nodes
$d(G)$ – the diameter of a connected graph G
E_{ij} – highlighted elements of CNI
$EF_j(t)$ – the distribution function (unreliability function) of the critical network infrastructure arc E_j
E_i – system components
E_{i_e} – arcs (edges) of the CNI
$e_i(t)$ – reliability state of element $E_i$at time t, $t \geq 0$
$e_{i_e}^{bin}$ – a binary random variable describing the state of the edge/arc of CNI E_{i_e}
$ER_j(t)$ – the reliability function of the two-state arc of the CNI E_j
ET_j – independent random variables describing the lifetimes of two-state arcs of the critical network infrastructure E_j
$ET_j(u)$ – independent random variables representing the lifetimes of the CNI arcs in the subset of reliability states
$F_{ij}(t)$ – distribution of lifetimes T_{ij} of the network components E_{ij}
$F_i(t)$ – distribution function (unreliability function) of element E_i
$F(t)$ – distribution function (unreliability function) of the system/CNI
$\varphi_{C-k|n:F}(\cdot)$ – a structural function of the "consecutive k out of n: F" CNI
$\varphi_{k|n}(\cdot)$ – a structural function of the "k out of n" CNI
$\varphi_p(\cdot)$ – a structural function of the parallel CNI
$\varphi_s(\cdot)$ – a structural function of the serial CNI
$\gamma(G)$ – domination number
$\gamma_C(G)$ – connected domination number
$\gamma_I(G)$ – independent domination number
$\gamma_{Iw}(G)$ – weighted independent domination number
$\gamma'(G)$ – edge-domination number
K_n – a complete graph with n vertices
$\xi(G)$ – subdivision number
m – a number of edges in graph/the number of edges/arcs of the CNI
n – a number of system's components/number of vertices in graph/number of vertices/nodes of the CNI
$NT(u)$ – an independent random variable representing the network's time in the subset of reliability states
P_n – a simple path of length n
$R_i(t))$ – reliability function of the two-state element E_i
$R_{ij}(t)$ – reliability function of network elements of critical infrastructures E_{ij}
$R_K(G)$ – the reliability of a graph/network G with distinguished K nodes

$R(t)$ – reliability function of the two-state system/CNI
$s(t)$ – reliability state of the system at time t
$s_{e_j}(t)$ – reliability state of edge E_j at time t
$s_{NT}(t)$ – reliability state of the CNI at time t
$s_{v_i}(t)$ – reliability state of vertex element V_i at time t
σ_i – the mixed strategy of player i
T – an independent random variable describing the uptime of a two-state complex technical system
T_i – independent random variables describing the lifetimes of two-state system elements E_i
T_{ij} – independent random variables representing the lifespan of components E_{ij}
T_n – a tree with n vertices
$T_i(u)$ – independent random variables representing the lifetimes of the system components in the subset of reliability states
u_i – payoff function of $i-th$ player
$v_{i_v}^{bin}$ – a binary random variable describing the state of the nodes/vertices of the CNI V_{i_v}
w_{ij} – payoff value for strategy a_i
$VF_i(t)$ – the distribution function (reliability function) of the vertices V_i
V_{i_v} – nodes/vertices of the CNI
$VR_i(t)$ – the reliability function of the two-state network nodes V_i
VT_i – independent random variables describing the lifetimes of two-state nodes (vertices) of the system network V_i

Author

Dr. Sambor Guze, professor at the Gdynia Maritime University, is a graduate of the Faculty of Mathematics, Physics and Computer Science at the University of Gdańsk, where he completed his Master's degree in Mathematics, specialising in Computer Science, in 2004. He obtained further academic degrees from the Maritime University of Szczecin, earning a PhD in Transport in 2011 and a postdoctoral degree (habilitation) in Civil Engineering and Transport in 2020.

From the beginning of his research and teaching career, he has been associated with the Department of Mathematics and the Faculty of Navigation, where he began his employment as an assistant in 2005. He is currently employed as a professor.

He is the author or co-author of over 50 publications in scientific journals, monographs, and peer-reviewed conference proceedings, as well as an editor and reviewer. He was an assistant supervisor for two completed doctoral dissertations.

He serves on the program and organising committees of national and international scientific conferences, and is a member of national and international scientific organisations such as the Polish Mathematical Society (PTM), the Polish Society for Safety and Reliability (PTBN), the Polish Navigation Forum (PFN), the Polish Society for Operational and Systems Research (PTBOiS), and the European Safety and Reliability Association (ESRA). Dr. Sambor Guze's research focuses on graph theory, optimisation, risk analysis, system safety and security, transportation networks and processes, and multi-criteria decision analysis.

List of Figures

List of Tables

1 Introduction to Critical Infrastructures Safety, Reliability, and Vulnerability

According to the definition, critical infrastructures are responsible for the reliable functioning of societies. Some of the most important are transport, energy, and information networks. This chapter discusses the basic concepts related to the safety, reliability, and vulnerability of critical infrastructures, particularly critical network infrastructures. It is the basis for the following three chapters, providing a canvas for presenting models and algorithms introduced in them.

1.1 GENERAL APPROACH AND BASIC NOTATIONS

The beginning of the 21st century is full of events that fully demonstrate what is essential for the proper and uninterrupted functioning of societies and countries. The challenge for today's national economies and societies is to ensure continuous access to basic services, products, and technologies. The end of the 20th century and the beginning of the 21st century have been and still are full of events that have negatively impacted our daily lives. First of all, terrorism, wars, and pandemics should be mentioned here – these are undesirable events, "the sources of which are primarily human activities. On the other hand, there are threats of natural origin – floods, earthquakes, volcanic eruptions, tornadoes, tsunamis, mudslides, etc. Natural hazards will intensify due to climate change. These changes depend on many factors, but primarily on the quantity of greenhouse gases in the atmosphere, ocean currents, carbon dioxide content, plate tectonics, volcanic eruptions, changes in the Earth's orbit, axial tilt and precession, the strength of the Sun, meteorite impacts, and changes in land cover.

Since we must learn to live in a changing natural and geopolitical environment, this is precisely why most countries around the world have defined so-called *critical infrastructures (CIs)* . In the European Union, their definition is as follows:

"Critical Infrastructure (CI) is an asset, system, or part thereof located in Member States that is essential for the maintenance of vital societal functions, health, safety, security, economic, or social well-being of people, and the disruption or

DOI: 10.1201/9781003661450-1

destruction of which would have a significant impact on a Member State as a result of the failure to maintain those functions."

In the United States, CIs first appeared in 1996 in Executive Order 13010. In 1998, the "cyber" part was added to Presidential Decision Directive 63 (PDD-63). This was additionally because the main aim of PDD-63 was to establish and develop a capacity to protect the CIs. The current definition in the United States is written in Section 1016(e) of the USA Patriot Act of 2001 and is as follows:

"Critical infrastructures are systems and goods, both physical and virtual, so vital to the nation that their malfunctioning or destruction would produce a debilitating impact on the security of citizens, the economic security of the nation, national public health, and any combination thereof."

In the United Kingdom, CI is called national infrastructure. These are the facilities, systems, sites, information, people, networks, and processes necessary for a country to function and upon which daily life depends. It also includes some functions, sites, and organisations that are not critical to the maintenance of essential services, but which require protection due to the potential danger to the public (civilian nuclear and chemical sites, for example). Technological advancements have forced legislators to include cybersecurity issues in their definitions. Depending on the country, the number of so-called critical sectors varies. A detailed list of CI (sectors) for the European Union, United States, United Kingdom, and Poland is presented in Table 1.1.

The data in Table 1.1 demonstrate differences in the identification of infrastructure, or more broadly, critical sectors. Poland, of course, identified and designated its infrastructure in accordance with the EU directive. Nevertheless, it is clear how a legal framework can be further refined to suit the needs of a given member state. Other differences between the European Union, United Kingdom, United States, and Poland likely stem from the experiences of individual countries in their daily operations, the impact of individual sectors on the economy, and the complexity of these sectors. For example, consider the dams critical sector identified in the US system. One could imagine them being in either the energy sector, the food and agriculture sector, or the water and wastewater sector. However, their importance for ensuring the country's daily security requires treating the dam system as a separate CI.

To better understand the definition of CI, it is necessary to clarify the basic concepts related to it. The most minor component of CI is an asset. It is any person, facility, material, information, or activity with a positive value to a system sector. The asset may have value to an adversary and an owner, although the nature and magnitude of these values may differ. Assets may be categorised in many ways, including people, information, equipment, facilities, and activities or operations.

Table 1.1
Critical Infrastructures in the European Union, United Kingdom, United States, and Poland

EU	UK	USA	Poland
Banking	Chemicals	Chemical	Communications
Digital infrastructure	Civil nuclear	Commercial facilities	Emergency services
Drinking water	Communications	Communications	Energy, energy raw materials, and fuel supply
Financial market infrastructure	Defence	Critical manufacturing	Financial systems
Energy	Emergency services	Dams	Food supply
Health	Energy	Defence industrial base	Healthcare
Production, processing and distribution of food	Finance	Emergency services	IT networks
Public administration	Food	Energy	Production, storage, and handling of chemical and radioactive substances, including hazardous substance pipelines
Space	Government	Financial services	Public administration continuity
Transport	Health	Food and agriculture	Transport
Wastewater	Space	government services and facilities	Water supply
	Transport	healthcare and public health	
	Water	information technology	
		Nuclear reactors, materials, and waste	
		transportation systems	
		Water and wastewater	

Source: Own work based on https://www.gov.pl/web/rcb-en/critical-infrastructure-systems https://eur-lex.europa.eu/legal-content/EN/TXT/?uri=CELEX:32022L2557; https://www.gov.uk/government/organisations/national-infrastructure-and-service-transformation-authority; https://www.cisa.gov/topics/critical-infrastructure-security-and-resilience/critical-infrastructure-sectors

Defining and designating critical sectors is the first step in ensuring the continuity of the daily functioning of economies and societies. However, more important is the protection of these designated sectors, especially their components. Therefore, each country designating CI implements plans for its protection.

When analysing the selected critical sectors or infrastructures, we must take into account dependency and interdependency. The first concept describes a unidirectional relationship between two assets (e.g., CI, firm, organisation, or facility), where the operations of the first asset affect the operations of the second asset. Interdependency is a mutually reliant relationship between entities (assets, objects, individuals, or groups). An important thing is that the interdependency does not need to be equal in both directions.

Simply identifying and designating national CIs is not sufficient. The next step is always to implement plans to protect these infrastructures, or more broadly, sectors. Here, we briefly focus on EU legislation in this area.

In June 2004, the European Council initiated the preparation of a strategy for protecting CIs within the Union's territory from potential terrorist attacks, which led the Commission to issue Communication 702/2004. In 2008, the Commission's activity led to the approval of Directive 2008/114/EC, which currently forms the basis of EU legislation on CI. The Directive aims to establish an EU process for identifying and designating European Critical Infrastructure (ECI) and to develop an approach to enhancing their protection. To this end, EU countries must use cross-cutting criteria, such as potential losses, economic effects, and impacts on citizens, as well as sectoral criteria specific to the type of ECI.

EU countries regularly review the identification and designation of ECIs. Thus, EU regulations took into account the relationship between CI and information systems. This was reflected in the fact that both the concepts and the regulatory method of the new Critical Entities Resilience (CER) Directive were modelled on the solutions of the previously introduced Directive 2016/1148 on cybersecurity (hereinafter referred to as the NIS Directive). In December 2022, a new directive on CI protection was adopted – Directive (EU) 2022/2557 of the European Parliament and of the Council of December 14, 2022 on the resilience of critical entities and repealing Council Directive 2008/114/EC (the CER Directive). This Directive was issued alongside the new EU Directive 2022/2555 on cybersecurity (the so-called NIS2 Directive) and the new law on the digital operational resilience of the financial sector (the Digital Operational Resilience Act). It is worth noting that both the EU Security Strategy for 2020–2025 and the new 2020 Counter-Terrorism Programme, adopted by the European Commission, emphasise the importance of CI resilience against digital threats. CI protection was also included in the EU Cybersecurity Strategy presented in December 2020. The new CER Directive is intended to replace the ECI Directive. At the same time, it significantly expands the scope of CI protection, introducing a new method of protecting these systems. It assumes increasing the resilience of specific categories of entities that operate in sectors distinguished in the Directive, i.e., in some areas of society's functioning, such as energy, transport, banking, and public administration. Ultimately, based on this Directive, Member States are required to identify so-called critical entities that provide essential services, which in turn play a fundamental role in maintaining crucial social functions, including important economic activities and functions related to public health, safety, and environmental protection.

In general, the CI protection aims to develop mechanisms to prevent threats not only related to terrorism but also to criminal activity, natural disasters, and other causes of accidents, based on the concept of comprehensive security. In particular, it plays a crucial role in preventing:

- catastrophic health effects and mass casualties
- impairment of public departments and agencies' ability to perform their essential missions or to ensure public health and safety
- the undermining of public administration's capacity to maintain order and deliver the minimum essential public services
- damage to the private sector's ability to ensure the orderly functioning of the economy and the delivery of essential services
- harm to the economy through the cascading disruption of other critical CI and key resources.

Therefore, it is necessary to develop tools that enable the efficient and effective creation of plans to protect not only the infrastructure itself but also the services associated with it.

1.2 CRITICAL NETWORK INFRASTRUCTURES

The literature on the subject indicates that, of the sectors shown in Table 1.1, the most critical is the energy sector. All others will not function without electricity. The COVID-19 pandemic, in turn, has highlighted the importance of maintaining the smooth operation of transport, both globally and, above all, locally. Furthermore, the geopolitical situation and the hybrid war that has been raging for over a decade demonstrate the importance of the IT sector and ICT and telecommunications networks. Thus, it is important to focus on these three CIs, which are responsible for the reliable flow of goods, people, information, and energy in the above-mentioned groups of CI.

CRITICAL NETWORK INFRASTRUCTURES

The transport networks (roads, rail, water, air, pipelines), telecommunications, and electricity transmission networks are referred to as the critical network infrastructures (CNIs).

1.2.1 TRANSPORTATION NETWORKS

According to the literature (Cascetta, 2001; Rodrigue et al., 2017), a transport system is defined as a combination of elements such as means of transport, infrastructure, human resources, and their mutual interactions, which serve to generate demand for travel in a given area and satisfy it by providing transport services. The above definition is very general, but simultaneously easy to adapt to various modes of transport. It allows the use of a structure of connections between transport system elements that are adapted to the problem being solved. It was assumed that the best and most natural way to record transport systems is a network representing the infrastructure of this system and providing the possibility of describing the flow of traffic occurring in it (Newell, 1980; Rodrigue et al., 2017). This should be agreed upon because the structure and traffic flows in the transport network best reflect the dependencies occurring between individual elements of the transport system and thus determine it. The following three main types of transport networks can be distinguished (Jasny et al., 2009; Rodrigue et al., 2017; Solé and Valverde, 2004):

Regular networks are those in which all vertices have the same degree, indicating the number of incident edges. This regularity suggests a high level of spatial organisation, as observed in road networks (Solé and Valverde, 2004).

Small-world networks feature dense connections among proximal nodes while maintaining a limited set of crucial connections with distant nodes. Such networks are highly sensitive to failures at nodes with many close neighbours, often referred to as hubs (Cui et al., 2007; Solé and Valverde, 2004; Watts and Strogatz, 1998).

Scale-free networks are characterised by a hierarchical arrangement with certain vertices featuring many neighbours and others featuring only a few (low vertex degree) . These networks evolve easily as new nodes tend to connect with larger nodes (Barab´asi, 2009; Barab´asi and Albert, 1999; Solé and Valverde, 2004).

A very important type of transportation network is the pipeline network. Their daily functioning is evident to us, and we become interested in them when they stop functioning correctly. However, they are complex systems tasked with transporting essential resources such as water, oil, and gas over vast distances. In doing so, they supply homes, businesses, and industry with basic utilities. Therefore, pipeline networks play a crucial role in ensuring the efficient and safe distribution of all resources necessary for the functioning of countries and societies.

These networks consist of interconnected pipes, fittings (valves and joints), pumps (compressors or blowers in the case of gas pipelines), booster stations (i.e., intermediate pumping stations placed along the pipeline to house pumps or compressors), storage facilities connected to the pipe, intake and outlet structures, control systems (i.e., flowmeters and other sensors), and a communication system (e.g., uses microwaves, cables, and satellites). They work together to move fluids and gases from source to destination. The design and maintenance processes of pipeline networks require careful planning and ongoing monitoring to ensure reliability and safety. These networks are always the first strategic goal for terrorism or war. As a result, pipeline operators face numerous challenges in keeping these critical systems running smoothly. It ranges from daily maintenance (e.g., corrosion control) to ensuring their integrity (e.g., after a natural disaster, malfunction, or terrorist attack).

Nowadays, long-distance pipelines are mainly automated. Monitoring and management are done at the headquarters of the pipeline company. The computer monitors pressure, flow rates, and other parameters at various locations along the pipe, performs many on-line computations. It allows to send commands to the field to control the operation of the valves and pumps. Manual intervention is frequently needed to modify automatic operation, as when different batches of fluids are directed to different temporary storage tanks. The same approach applies when the operator of the pipeline must shut down or restart the system.

Due to the nature of pipeline design and construction, there are no specific types of operation for them, as is the case with transportation networks. Their shape depends on the designated pipeline route, source, and destination.

1.2.2 TELECOMMUNICATION NETWORKS

According to the literature, a telecommunication network comprises a group of terminal nodes, intermediate nodes, and links that are interconnected to facilitate telecommunication between terminals. These transmission links connect the nodes. In this way of thinking, telecommunication networks are intricate systems of links (such as landlines, fibre optics, and 5G technology) and nodes (like routers, switches, and servers). The main functionality is to seamlessly transmit messages from node A to node B, regardless of network distance. The nodes utilise different types of switching, such as circuit, message, or packet switching, to pass the signal through the appropriate links and nodes to reach the intended destination terminal. Examples of telecommunications networks include computer networks, the Internet, the public switched telephone network), the global Telex network, the aeronautical ACARS network, and the wireless radio networks of cell phone telecommunication providers.

The various topologies are as follows:

Star Network uses a central computer connected to different small computers.

Bus Network uses one circuit to link other computers.

Ring Network is the most independent kind of network and does not rely on a central host computer.

Mesh Network is one in which each node is connected to the others and can send and receive data as well as relay information from other nodes.

Tree Network is an extension of the bus topology, combining the star and bus topologies.

Hybrid Network combines elements of many different topologies.

Depending on the range of operation of the telecommunications network, we distinguish the following types:

Local Area Network (LAN) is designed for small areas like an office, a group of buildings, or a factory.

Wide Area Network (WAN) can be private or public, used for networks that cover large distances, such as those spanning multiple states within a country. They can also be used to cover a large city or metropolitan area.
Metropolitan Area Network (MAN) is quite a bigger version of LAN and designed to extend over the whole city. Usually, it connects several LANs into the bigger network.

1.2.3 ENERGY NETWORKS

An electric power system can schematically be divided into generation units (generators, transformers etc.), delivery systems, and users. Mainly, the electric power delivery system consists of:

Transmission grids (high-voltage) are meshed networks that connect large generating stations (e.g., hydropower and nuclear power), sub-transmission grids, and very large users. The main operational aim of transmission grids is to enable power trading with other countries and to facilitate the optimisation of generation within a country.
Sub-transmission grids, or regional grids, are radial or locally meshed networks connected to the transmission grid via infeed points. Smaller generating plants (e.g., wind power stations and gas turbines) and large users are connected to these grids.
Distribution grids (low-voltage) are radial networks that carry the electric power from the higher voltage levels to the final users. The number of levels in a distribution grid depends upon the density and magnitude of demand and the terrain.

The topology of energy networks for sub-transmission grids is as follows:

A *radial network* is the most classic and intuitive form of power supply – electricity flows in one, unchanging direction: from the power supply station towards the end user.
A *closed-loop network* elegantly combines the simplicity of a radial network with the robustness of mesh systems. Its foundation lies in the ability to power a given point from two different directions, with only one side receiving power under normal conditions. In contrast, the other side remains as a reserve (ready for switching).
A mesh network is the most advanced and complex distribution topology, in which medium-voltage lines form a dense network of interconnections. Current can flow along multiple possible paths, and each customer can be supplied from several different directions simultaneously.

1.3 VULNERABILITY AND SAFETY OF CRITICAL INFRASTRUCTURES

Generally, Holmgren and Molin (2006) introduce the following vulnerability working definition:

"Vulnerability is the collection of properties of an infrastructure system that might weaken or limit its ability to maintain its intended function or provide its intended services when exposed to threats and hazards that originate both within and outside of the boundaries of the system."

In simpler words, the term *vulnerability* means a physical characteristic or operational attribute that makes an entity open to exploitation or susceptible to a given threat (Appert and Chapelon, 2007; Berdica and Eliasson, 2004; D'Este and Taylor, 2003; Du and Nicholson, 1997; Einarsson and Rausand, 1998; Holmgren, 2004a; Jenelius et al., 2006; Proag, 2014; Reggiani et al., 2015; Sarewitz et al., 2003; Taylor and D'Este, 2004).

In terms of reliability, it can be measured as the probability that a system will transition to a critical state or worse within a shorter time than expected. This may be determined by certain external factors that cause significant negative impacts on other vulnerable systems (consequences above a constant, predetermined level). In other words, this approach can be expressed as the inherent properties of a system or its components, assets, communities, and the environment that make them susceptible to the adverse effects of natural and anthropogenic hazards.

In the language of societal consequences, Holmgren (2007) proposed the following formula describes a vulnerability of system. Let $C(t)$ be the societal consequences of a disturbance that occurs at time $t \in T$ (T is given period of time) and c is a critical value of consequences. Under these assumptions, the vulnerability of the infrastructure system can be calculated by

$$P(\max_{t \in T} C(t) > c) \tag{1.1}$$

Next, it is possible to extend the formula 1.1 to conditional approach. Firstly, let A_i is an initiating event, then the conditional vulnerability is given (Holmgren, 2007)

$$P(\max_{t \in T} C(t) > c | A_i) \tag{1.2}$$

Finally, the initiating event can occur either due to failure of technical components or other cause. The first one gives us opportunity to estimate failure rate (frequency). The second has to be calculate based on Eq. 1.1. Thus, the total probability is given by:

$$P(\max_{t \in T} C(t) > c) = \sum P(A_i) \cdot P(\max_{t \in T} C(t) > c | A_i) \tag{1.3}$$

A concept closely related to the vulnerability is *risk*. In the literature, it is defined as the combination of consequences and the probability of their occurrence. It is crucial to ensure the security of CI. Risk assessment is a basic tool for vulnerabilities identification, potential consequences estimation, and mitigation measures development. The steps of this approach are generally well-known (e.g., Šarūnienė et al., 2024) and common for all types of objects and are as follows:

1. system description
2. hazard identification
3. probabilities and consequences estimation and identification
4. risk evaluation

First step of the risk assessment needs information about a CI's detailed description. We have to identify its limits (functional and structural), dependencies, and interdependencies. This is some kind of model of considered CI.

Next, we have to identify the threats and hazards influencing CI. One possible categorisation of these hazards can be as follows (Šarūnienė et al., 2024):

1. Natural, which include adverse and extreme natural phenomena, such as hazards occurring in the air, on the ground, and under the ground, as well as external fires in proximity to CI. They are the most common type of hazards and depend on geographic localisation.
2. Technical, which include the unreliable operation of CI resulting from various accidents and failures that occur due to technical reasons.
3. Economic, which include various risks related to economic factors, i.e. economic crises, the isolation of CI, or the dominance of a particular fuel source, producer, or supplier in the energy CI, etc.
4. Socio- and geopolitical, which are crucial to identify due to their potential for severe consequences. This type of hazard significantly influences decision-making processes regarding CI development; here, it should also include cyberattacks and terrorism.

Certainly, the man-made hazards (or human caused) are either non-malicious or malicious. The first group includes, among others, transport accidents, emergency events on industrial and military facilities, loss of CI, and many others. The second group comprises terrorism, cyberattacks, diversions, and sabotage.

Third step of risk assessment is about probabilities evaluation and consequences analysis. The probability can be find according to chain rule, i.e. conditional probability (if events are dependent). For given events $A_1, \ldots, A_n$, and their intersection has not probability zero, we have the following formula:

$$P(A_1, \ldots, A_n) = \prod_{k=1}^{n} P(A_k | A_1 \cap A_2 \cap \ldots \cap A_{k-1}) \tag{1.4}$$

Eventually, if components of the CI are independent, we can use joint distribution for independent variables. Two random variables X and Y are independent if and only if the joint cumulative distribution function satisfies the following formula:

$$F_{X,Y}(x,y) = F_X(x) \cdot F_Y(y) \tag{1.5}$$

The concept of vulnerability is naturally related to the idea of the *resilience* of a system or network. The essence of this concept is the internal ability of a system (network) to maintain or recover a dynamically stable state that allows it to continue operations after a serious incident caused by natural or man-made factors, while continuing to occur (Hollnagel et al., 2006; Patriarca et al., 2018). Hence, the resilience of a CNI can be defined as its ability to resume operations at a level similar to that before the disruption occurred at nodes, arcs, or both. The fewer disruptions in terms of capacity and fluidity, and the faster the network resumes regular operation, the greater its resilience. In other words, a key measure of CI resilience is the ability and time required to mitigate the consequences of the above-mentioned groups of threats materialising. The resilience of any network is influenced mainly by its structure, particularly its redundancy.

As mentioned earlier, the primary task of CIs is to ensure the uninterrupted operation of particular services for the benefit of society. Therefore, the ability to mitigate effects, as well as to proactively increase the resilience of CI, strengthens their performance. This ultimately ensures their continued operation, known as business continuity. This approach is reflected in the development of CI protection programs mentioned in Section 1.1.

1.4 PRINCIPLES OF RELIABILITY

The basic assumption is that CIs or CNIs are complex socio-technical systems. Therefore, based on systems theory, we can use modelling and analysis methods designed for complex technical systems, particularly when dealing with CNIs. Thus, we use the well-known theory introduced in works such as Blokus-Roszkowska (2016); Guze (2009); Kołowrocki (2004); Khan (2014); Kołowrocki and Soszyńska-Budny (2011); Mine and Kawai (1982), and Nowakowski (2011).

Generally, *reliability* is a characteristic of a technical system that describes its ability to perform specific tasks without interruption. In mathematical terms, it is a probabilistic measure of a system or its component, describing its ability to operate continuously and without interruption. The theory of technical system reliability distinguishes between a structural and a functional approach. The former allows for modelling and analysis of the structure of a specific system. This requires distinguishing and naming elements, as well as the connections between them. It directly impacts the behaviour of a given system due to its structure. This is typical of CIs, which are characterised by interconnections within the specific infrastructure and with other infrastructures.

In terms of reliability, technical systems can be either two-state or multi-state. The first type refers to systems that can be either operational or non-operational. In the case of multi-state systems , however, they or their components reside in a specific set of reliability states, e.g., $\{0,1,\dots,z\}$. The states are arranged from state 0, which is the worst, to state z, which is the best. The two-state model of CI can also be represented as the set $\{0,1\}$, where $z=1$.

For both approaches to technical systems reliability analysis, an important concept is the lifetime of a system or component. This is the time during which the system or component operates correctly. When they lose their ability to perform the dedicated function, then it is a failure. The rate at which failures occur over time is referred to as the failure rate. An important parameter is the *mean time to failure*

(MTTF), which means the average time a system is expected to function before a failure.

When we consider the CNI (introduce in Section 1.2), we interest in the network reliability. One of the first works on network reliability was a paper by Rosenthal (1977), in which the author noted, among other things, that the problem of finding a fault tree and testing network reliability is difficult, and finding a fast algorithm to solve these problems is also not easy.

A review of the state of the art in this area can be found in the paper by Agrawal and Barlow (1984). The authors also noted the connection between network reliability and the more general problem of computing the reliability of coherent structures, introduced by Barlow and Proschan (1975). The concept of reliability of coherent structures in technical systems means that the system is represented by a structural function and that each of its elements is essential to its reliability. Agrawal and Barlow Willie (1978) pointed out that coherent structures include both directed and undirected networks and logical trees. Furthermore, they defined the problem of reliability of a transportation network as calculating the probability that two network nodes are connected. At the same time, possible connections between two selected nodes each create different sets of shortest paths in the network (Agrawal and Barlow, 1984). They identified three methods for determining reliability based on the minimal path set and the minimal cutting set. The first term is understood as the smallest set of network elements (edges) whose operation indicates that a path exists between the distinguished nodes. The operation of a component means that travel along the section of the road represented by these edges is possible. A minimal cutting set, on the other hand, is the smallest set of elements (edges) whose inaction (e.g., blocking a road) results in no connections between selected network nodes. However, these sets are not the only way to calculate the reliability of a network represented by a graph.

Assuming that $R_K(G)$ denotes the reliability of a graph/network G with distinguished K nodes, expressed as the probability that the elements of at least one smallest set of paths are healthy (movement is possible in the transport network) or expressed in the dual language of cutting sets, they can be defined as one minus the probability of the event that all elements of at least one cutting set are damaged (movement is impossible in the transport network). Additionally, assuming that p denotes the number of sets of shortest paths, A_i denotes the event that all elements in the i-th smallest set of paths are healthy, and $\overline{A_i}$ denotes the complementary event of A_i, the network reliability can be expressed using (Agrawal and Barlow, 1984):

a) inclusion-exclusion method

$$R_K(G) = \sum_{i=1}^{p} P(A_i) - \sum_{i=1}^{n} \sum_{j<i} P[A_i A_j] + \cdots + (-1)^{p-1} P[A_1 A_2 \cdots A_p] \quad (1.6)$$

b) sum of disjoint product

$$R_K(G) = P(A_1) + P(\overline{A_1} A_2) + \cdots + P(\overline{A_1 A_2 \cdots A_{p-1}} A_p) \quad (1.7)$$

c) pivotal decomposition/factoring

$$R_K(G) = p_e R_K(G|e) + (1 - p_e) R_K(G|\bar{e}) \tag{1.8}$$

where

$R_K(G|e)$ expresses the reliability of the graph G under the condition that edge e is functional

$R_K(G|\bar{e})$ is the reliability of the graph G under the condition that edge e is not functional

It allows for the use of two approaches to reliability modelling in analyses related to systems, depending on the needs. The more straightforward approach is the two-state approach, meaning the system operates in a binary manner. The multi-state approach is more difficult, but closer to the actual description. In the latter case, we assume that the network is subject to an ageing process and its reliability deteriorates. Such a network/system may also be subject to renewal. In research (e.g., Kołowrocki, 2004, 2013; Khan, 2014; Kołowrocki and Soszyńska-Budny, 2011) on the reliability of complex technical systems, the following notations are used:

- n – number of system's components;
- E_i, $i = 1,2,\ldots,n$ – system components; in the case of the two-state approach, all system elements, as well as the system itself, can be functional or non-functional, whereas in the case of the multi-state approach, all system elements, as well as the system itself, can be in a set of reliability states $\{0,1,\ldots,z\}$, where the states are arranged from state 0 – the worst to state z, which is the best;
- T_i – independent random variables describing the lifetimes of two-state system elements E_i, $i = 1,2,\ldots,n$;
- T – an independent random variable describing the uptime of a two-state complex technical system;
- $T_i(u)$, $i = 1,2,\ldots,n$ – independent random variables representing the lifetimes of the system components in the subset of reliability states $\{u, u+1,\ldots,z\}$, when they were in state z at time t = 0;
- $T(u)$ – a random variable representing the time the system stays in the subset of reliability states $\{u, u+1,\ldots,z\}$, assuming that it was in state z at time t = 0, the system condition deteriorates at time t without repair;
- $R_i(t) = P(T_i > t)$, $t \in <0, +\infty)$ – reliability function of the two-state element E_i, $i = 1,2,\ldots,n$;
- $F_i(t) = 1 - R_i(t) = P(T_i \leq t)$, $t \in <0, +\infty)$ – distribution function (unreliability function) of element E_i, $i = 1,2,\ldots,n$;
- $R(t) = P(T > t)$, $t \in <0, +\infty)$ – reliability function of the two-state system;
- $F(t) = 1 - R(t) = P(T \leq t)$, $t \in <0, +\infty)$ – distribution function (unreliability function) of the system;
- $e_i(t)$ – reliability state of element $E_i$at time t, $t \geq 0$;
- $s(t)$ – reliability state of the system at time t, $t \geq 0$.

1.4.1 TWO-STATE RELIABILITY APPROACH

Taking into account the above notations, the individual reliability structures in the case of the two-state approach are defined as follows. A two-state complex technical system is called

- serial if the relationship expresses its lifetimes T is in following form:

$$T = \min_{1 \le i \le n} \{T_i\} \tag{1.9}$$

- parallel if the relationship expresses its lifetimes T is in following form:

$$T = \max_{1 \le i \le n} \{T_i\} \tag{1.10}$$

- a "m out of n" if its lifetime T is is in following form::

$$T = T_{(n-m+1)}, m = 1, 2, \ldots, n, \tag{1.11}$$

  where

  $T_{(n-m+1)}$ is the m-th statistic of the highest degree in the sequence of lifetimes of the system components $T_1, T_2, \ldots, T_n$.
- a "consecutive k out of n: F" if it is faulty if and only if at least k of its consecutively ordered elements $E_1, E_2, \ldots, E_n$ are faulty.

In other words, a two-state, serial technical system will be efficient if all of its components are efficient. A parallel technical system, on the other hand, will be efficient if at least one of its components is efficient. In the case of a threshold system, it is efficient if its m out of n available components are efficient. In the special case where m=n, we have a parallel system, and when m=1, we have a serial system. With respect to the defined structures of two-state systems, the reliability functions – assuming that failures are independent events – are given by the following formulas:

- serial systems:

$$R(t) = \prod_{i=1}^{n} R_i(t) \tag{1.12}$$

  where $t \in < 0, +\infty)$
- parallel systems:

$$R(t) = 1 - \prod_{i=1}^{n} F_i(t) \tag{1.13}$$

  where $t \in < 0, +\infty)$
- "m out of n" systems:

$$R(t) = 1 - \sum_{\substack{r_1, r_2, \ldots, r_n = 0 \\ r_1 + r_2 + \ldots + r_n \le m-1}}^{1} [R_i(t)]^{r_i} [F_i(t)]^{1-r_i} \tag{1.14}$$

or

$$R(t) = \sum_{\substack{r_1, r_2, \ldots, r_n = 0 \\ r_1 + r_2 + \ldots + r_n \leq \bar{m}}}^{1} [F_i(t)]^{r_i} [R_i(t)]^{1-r_i} \tag{1.15}$$

where

$t \in < 0, +\infty)$

$\bar{m} = n - m$

- "consecutive k out of n: F" systems:

$$R(t) = R_{k,n}(t) \tag{1.16}$$

where $R_{k,n}(t)$ is given by the following recursive formula:

$$R_{k,l}(t) = \begin{cases} 1 - \prod_{j=1}^{l} F_j(t) & \text{for } k = l, \\ R_l(t) R_{k,l-1}(t) & \\ + \sum_{i=1}^{k-1} R_{l-i}(t) R_{k,l-i-1}(t) \cdot \prod_{j=l-i+1}^{l} F_j(t) & \text{for } k < l, \end{cases} \tag{1.17}$$

for $l = 1, 2, \ldots, n$, $t \in < 0, +\infty)$.

In the case of two-state systems, the following parameters can also be determined to characterise them (Kołowrocki, 2004, 2013; Khan, 2014; Kołowrocki and Soszyńska-Budny, 2011):

- average system's lifetime

$$E[T] = \int_0^\infty R(t) dt \tag{1.18}$$

- second-order moment of the system's lifetime

$$E[T^2] = 2 \int_0^\infty t R(t) dt \tag{1.19}$$

- standard deviation of system lifetime

$$\sigma = \sqrt{D[T]} \tag{1.20}$$

where

$$D[T)] = E[T^2] - (E[T])^2. \tag{1.21}$$

1.4.2 MULTI-STATE RELIABILITY APPROACH

A more realistic modelling approach to the reliability of technical systems is the multi-state approach, as it allows for a more complete and realistic representation of the analysed system. However, it requires the clarification of certain concepts. Concerning the system and its components, we assume that $\{0, 1, \ldots, z\}$ is an ordered set of their states, where state 0 is considered the worst, and state z is the best, in terms of reliability. Additionally, the theory presented in works such as Kołowrocki (2004, 2013); Khan (2014); Kołowrocki and Soszyńska-Budny (2011), takes into account the fact that the system and its components, unless repaired, deteriorate over time.

With these assumptions in mind, vector functions of the components' and system's reliability have been defined in the literature on the reliability theory of multi-state complex technical systems. According to these assumptions, we have following formulae:

- the reliability function of elements E_i, $i = 1, 2, \ldots, n$ is described by the vector

$$R_i(t, \cdot) = [R_i(t, 0), R_i(t, 1), \ldots, R_i(t, z)] \tag{1.22}$$

  where

$$R_i(t, u) = P(e_i(t) \geq u | e_i(0) = z) \tag{1.23}$$

  is the probability of the event that element E_i, $i = 1, 2, \ldots, n$, at time $t, t \in < 0, +\infty)$ is in one of the states from the subset of states $\{u, u+1, \ldots, z\}$, while at time $t = 0$, it was in the best state z;
- the vector describes the system's reliability function:

$$\mathbf{R}(t, \cdot) = [\mathbf{R}(t, 0), \mathbf{R}(t, 1), \ldots, \mathbf{R}(t, z)] \tag{1.24}$$

  where

$$\mathbf{R}(t, u) = P(s(t) \geq u | s(0) = z) \tag{1.25}$$

  is the probability that the system at time t, $t \in < 0, \infty)$,is in one of the states from the subset of states $\{u, u+1, ..., z\}$, while at time $t = 0$, it was in the best state z.

The probability that element E_i, $i = 1, 2, \ldots, n$ is in state u, provided that at time $t = 0$ it was in state z, is given by the vector:

$$p_i(t, \cdot) = [p_i(t, 0), p_i(t, 1), \ldots, p_i(t, z)] \tag{1.26}$$

where

$$p_i(t, u) = P(e_i(t) = u | e_i(0) = z) \tag{1.27}$$

and $u = 1, 2, \ldots, z, , t \in < 0, +\infty)$.

However, the probability that the system is in state u, given that it was in state z at time $t = 0$, is given by the vector:

$$p(t,\cdot) = [p(t,0), p(t,1), \ldots, p(t,z)] \tag{1.28}$$

where

$$p(t,u) = P(s(t) = u|s(0) = z) \tag{1.29}$$

and $t \in < 0, +\infty)$, $u = 1, 2, \ldots, z$.

The introduction of the above concepts allows us to define the reliability structures of multi-state, ageing technical systems and their components, as well as the parameters characterising these systems. Analogously to the two-state system, the following definitions are true:

A multi-state technical system is

- serial if its time $T(u)$ in the subset of reliability states $\{u, u+1, \ldots, z\}$ is described by the relationship:

$$T(u) = \min_{1 \le i \le n} \{T_i(u)\} \tag{1.30}$$

where $u = 1, 2, ..., z$.

- parallel, if the time $T(u)$ of its stay in the subset of reliability states is described by the relationship:

$$T(u) = \max_{1 \le i \le n} \{T_i(u)\} \tag{1.31}$$

where $u = 1, 2, \ldots, z$.

- "m out of n" if the time $T(u)$ of its stay in the subset of reliability states $\{u, u+1, \ldots, z\}$ is described by the relationship:

$$T(u) = T_{(n-m+1)}(u) \tag{1.32}$$

where $T_{(n-m+1)}(u)$ is the m-th statistic of the highest degree in the sequence of the lifetimes of the elements $T_1(u)$, $T_2(u)$,..., $T_n(u)$, $u = 1, 2, \ldots, z$.

- "consecutive k of n: F," if it does not reside in the subset of reliability states $\{u, u+1, \ldots, z\}$, if and only if at least k of its n ordered elements $E_1, E_2, \ldots, E_n$ do not reside in this subset.

As in the case of a two-state system, we say that a system is in a subset of reliability states if all of its elements are in it. In the case of a parallel system, the system is in the subset of reliability states $\{u, u+1, \ldots, z\}$ if at least one of its elements is in it. In contrast, a multi-state, technical system of type "m out of n" is in the subset of reliability states $\{u, u+1, \ldots, z\}$ if at least m of its n elements are in this subset.

With respect to the reliability structures defined above, the coordinates of the reliability function vector are as follows:

- for a multi-state serial system:

$$\mathbf{R}(t,u) = \prod_{i=1}^{n} R_i(t,u) \tag{1.33}$$

where $t \in < 0, +\infty)$, $u = 1,2,\ldots,z$
- for a multi-state parallel system:

$$\mathbf{R}(t,u) = \prod_{i=1}^{n} F_i(t,u) \tag{1.34}$$

where $t \in < 0, +\infty)$, $u = 1,2,\ldots,z$,
- for a multi-state "m out of n" system:

$$\mathbf{R}(t,u) = 1 - \sum_{\substack{r_1,r_2,\ldots,r_n=0 \\ r_1+r_2+\ldots+r_n \le m-1}}^{1} [R_i(t,u)]^{r_i} [F_i(t,u)]^{1-r_i} \tag{1.35}$$

or

$$\mathbf{R}(t,u) = \sum_{\substack{r_1,r_2,\ldots,r_n=0 \\ r_1+r_2+\ldots+r_n \le \bar{m}}}^{1} [F_i(t,u)]^{r_i} [R_i(t,u)]^{1-r_i} \tag{1.36}$$

where $t \in < 0, +\infty)$, $\bar{m} = n - m$, $u = 1,2,\ldots,z$,
- for a multi-state "consecutive k out of n: F" system:

$$\mathbf{R}(t,u) = R_{k,n}(t,u) \tag{1.37}$$

where $R_{k,n}(t)$ is given by recursive formula:

$$R_{k,l}(t,u) = \begin{cases} 1 - \prod_{j=1}^{l} F_j(t,u) & \text{for } k = l, \\ R_l(t,u)R_{k,l-1}(t,u) & \\ + \sum_{i=1}^{k-1} R_{l-i}(t,u)R_{k,l-i-1}(t,u) \cdot \prod_{j=l-i+1}^{l} F_j(t,u) & \text{for } k < l, \end{cases} \tag{1.38}$$

where $l = 1,2,\ldots,n$, $t \in < 0, \infty$, $u = 1,2,\ldots,z$.

The basic parameters describing multi-state systems are most often given in terms of elements and the entire system. For element E_i, $i = 1,2,\ldots,n$, the selected parameters are given as follows (Kołowrocki, 2004, 2013; Khan, 2014; Kołowrocki and Soszyńska-Budny, 2011):

- the expected value of the residence time in the subset of states $\{u, u+1,\ldots,z\}$ is given by the formula:

$$M_i(u) = \int_0^{\infty} R_i(t,u)dt \tag{1.39}$$

- standard deviation of the residence time in the subset of states $\{u, u+1, \ldots, z\}$:

$$\sigma_i(u) = \sqrt{N_i(u) - [M_i(u)]^2} \tag{1.40}$$

where

$$N_i(u) = 2\int_0^\infty tR_i(t,u)dt, u = 1,2,\ldots,z. \tag{1.41}$$

The selected parameters for the entire system are as follows (Kołowrocki, 2004, 2013; Khan, 2014; Kołowrocki and Soszyńska-Budny, 2011):

- the expected value of the residence time in the subset of states $\{u, u+1, \ldots, z\}$ is given by the formula:

$$M(u) = \int_0^\infty \mathbf{R}(t,u)dt \tag{1.42}$$

- standard deviation of the residence time in the subset of states $\{u, u+1, \ldots, z\}$:

$$\sigma(u) = \sqrt{N(u) - [M(u)]^2} \tag{1.43}$$

where

$$N(u) = 2\int_0^\infty t\mathbf{R}(t,u)dt \tag{1.44}$$

Moreover, $\overline{M}_i(u) = \int_0^\infty p_i(t,u)dt$ and $M(u) = \int_0^\infty p(t,u)dt$ are the expected values of the residence time of the element E_i, $i = 1,2,\ldots,n$ and the system in state u, respectively, in the case when the above integrals exist.

1.4.3 RELIABILITY OF RENEWAL TECHNICAL SYSTEMS AND COMPONENTS

An extension of the reliability concept for ageing technical systems is the consideration of their *renewal*. It can either repair or replace a component with a new one. Renewal, in the sense of system operation, is a process that occurs over time and aims to restore a single component or the entire system to full operational efficiency. Renewal theory considers models with negligible and non-negligible renewal times.

Model with negligible renewal times

In the case of non-negligible renewal time, the renewal process is a sequence of moments when a given element is failure. In the analysis of the renewal process, the most important thing is to determine the number of failures (renewals) by time

t, denoted as $\nu(t)$. From a practical point of view, the best measure is the expected number of failures (renewals) by time t, called the *renewal function* $H(t)$:

$$H(t) = E[\nu(t)] = \sum nP_n(t) \tag{1.45}$$

where

$\nu(t)$ – a random variable representing the number of failures (renewals) until time t;
t_n – time until the n-th failure occurs;
$P_n(t) = P\{\nu(t) = n\}$.

The derivative of the renewal function is the renewal density $h(t)$ and it expresses the average number of failures in a small unit of time, starting from time t. The renewal density can be expressed as

$$h(t) = \sum_{n=1}^{\infty} f_n(t) \tag{1.46}$$

where

$f_n(t) = F_n'(t)$
$F_n(t) = P\{\nu(t) \geq n\} = P\{t_n < t\}$ – distribution function of a time to n-th failure
t_n – time until the n-th failure occurs.

The asymptotic behaviour of the renewal function, rather than its analytical form, has interesting applications. Using the asymptotic approach, the following inequality for renewal function is true

$$F(t) \leq H(T) \leq \frac{F(t)}{1 - F(t)} \tag{1.47}$$

and thus we get for $F(t) << 1$ the approximation

$$H(t) \approx F(t) \tag{1.48}$$

When we consider an ageing component (see Section 1.4.2), i.e. with a monotonically increasing intensity function, we get the next approximation

$$H(t) \leq \frac{t}{T_0} \tag{1.49}$$

Model with non-negligible renewal times

Availability describes the ability of a given component to remain in a working state and fulfil its functions at a given time or moment. It is defined as the probability that a given device/system will operate correctly when required. Availability is typically defined according to formula (Kołowrocki and Soszyńska-Budny, 2011):

$$A = \frac{E(T)}{E(T) + E(\Theta)} \tag{1.50}$$

where

> $E(T)$ – expected value of the random variable of the system/technical object's uptime
> $E(\Theta)$ – expected value of the random variable of the system/technical object's downtime (renewal and waiting for renewal)

For such defined values, the following can be determined:

- *mean time between failures (MTBF)*, which is defined as the average expected time between successive failures of a given system, assuming that the system undergoes alternating periods of failure and repair;
- *mean time to failure (MTTF)*, which is defined as the average time elapsed from the moment the system regains serviceability until a subsequent failure occurs;
- *mean time to repair (MTTR)*, which is the length of time during which a given system is down due to a failure;
- *mean downtime (MDT)* (interchangeably with MTTR), which includes the time of failure detection, its recognition and isolation, handling of the report, hiring a team to repair the failure, the time of its repair, and an operational test;
- *MUT,* which is defined as the average percentage of time a system is up and running;
- *MRT* is the average time of system repair (back to complete serviceability);
- *mean time to recovery (MTTV)*, which is the time at which the system regains its ability to function correctly. This moment occurs when the system recovers from the failure state due to the activation of protective or recovery mechanisms. The MTTV value is generally one or several orders of magnitude lower than the MTTR value, which translates into improved service readiness.

Assuming that the working time and downtime are known, availability can be calculated from the relationship:

$$A = \frac{MUT}{MUT + MDT} \tag{1.51}$$

From the operator's perspective, it's easier to measure MTBF and MTTR. When measured over a more extended period, MDT can approximate MTTR because the mean time to repair is equal to the downtime, assuming the downtime is caused by failures rather than maintenance. Similarly, the time between failures is the sum of downtime and operating time, so when measured over a sufficiently long period, the sum of MUT and MDT can approximate the MTBF value. Based on the above reasoning, availability can be expressed using the following formula (Nowakowski, 2011):

$$A = 1 - \frac{MTTR}{MTBF} \tag{1.52}$$

whereas unreadiness (U) is defined by the following formula:

$$U = 1 - A. \tag{1.53}$$

Availability and downtime are expressed as probabilities and range from 0 to 1. Availability typically ranges from 98% to 99.999%, which is difficult to compare. Therefore, downtime can be expressed in minutes per year. This expression defines the expected downtime of a device during a given operating period. This is another way of expressing availability. Downtime is more convenient and understandable than the concept of availability when assessing a given device or system. Typical values can be defined for various technical facilities/systems.

2 Modelling of Critical Network Infrastructures

This chapter presents well-known models and approaches to modelling critical network infrastructures, which provide the basis for introducing new tools for their modelling. The chapter introduces basic definitions and theorems related to graph theory, game theory, mathematical statistics, and the theory of stochastic processes. The chapter provides a basis for introducing new tools for analysing and optimising critical network infrastructures in the following chapters.

2.1 GENERAL APPROACH AND BASIC NOTATIONS FOR MODELLING

The main idea of modelling is the representation of some phenomenon or property we are trying to understand or study in terms of other phenomena we already understand. Modelling is done based on a model. It is an abstract or material structure, similar to the structure of the modelled object. This approach can provide information that meets the goals of research, design, or improvement of such an object. We build models for description, explanation, evaluation, decision, and forecasting. According to this, in engineering practice, we can highlight the following models: descriptive, physical, analogue, digital, abstract, and mathematical.

The simplest and at the same time fundamental division is between physical and abstract models – this is the classification by model type. Nevertheless, we can classify models by:

- modelling purpose
- level of abstraction
- representation type
- method of system description

In engineering, we can use different types of models, including structural, functional (or operational), and decisional models. Structural system (particular network) modelling involves abstractly representing physical objects – in our case, transportation, energy, and telecommunications networks – using nodes and edges. Nodes represent components and connections within the modelled network, such as a junction in a transportation network or a switch in a telecommunications system. Edges represent how network components are interconnected, for example, via power lines in power systems or fibre optics in telecommunications systems.

The functional model, in turn, represents the network's response to workloads, taking into account the physical flows within the network (energy, information,

DOI: 10.1201/9781003661450-2

vehicles, oil, water, etc.). Simultaneously, it allows for the calculation of their consequences for individual network components.

Finally, a decisional model represents the way of thinking when a decision has to be made. We can distinguish up to five decision-making models. The first is the *rational model*, which consists of six steps:

1. problem definition
2. defining criteria used to evaluate possible solutions
3. assigning weights to criteria
4. identifying options to choose from
5. evaluating options
6. determining the best solution.

The second is the model of *bounded rationality*. In this case, the decision-maker follows the rules of full rationality, but is not "perfect" in their tasks. Certain limitations prevent them from acting entirely rationally, such as cognitive and motivational factors.

The third type of decision-making model is the *heuristic* approach. It assumes a complete departure from full rationality and represents actual decision-making processes.

And the final possible model is the *dual-process* model. Its purpose is to integrate an approach focused on rational analysis with a strategy focused on intuition.

Taking into account the above mentioned models, we use graph theory as both structural and functional models of critical network infrastructures (CNIs). The game theory models proposed below are appropriate for modelling the functional and decisional side of CNIs. Both approaches can also be used to model the relationships between critical sectors and their dependencies and interdependencies.

2.1.1 GRAPH THEORY BASIC NOTATIONS

To represent CNIs, we most often consider simple, connected, undirected graphs. However, in some cases, this description is insufficient, and then it is necessary to use directed or weighted graphs. It will happen when the direction of traffic in the represented networks is taken into account, or some numbers can describe the network components. To make the following content understandable, we provide here the basic definitions of graph theory. They come from various sources, including Diestel (2000); Gross and Yellen (2004); Harrary (1969); Leeuwen (1986); Velammal (1997); Wilson (1998).

We assume that a pair of sets V and E represent a simple graph G, where V denotes a non-empty finite set of *vertices (nodes)* and E denotes the set of *edges (arcs)*, which is any subset of the set of unordered pairs of elements of V. Sometimes we will use $V(G)$ and $E(G)$ to denote the set of vertices and the set of edges of the graph G, respectively. We further assume that $|V| = |V(G)| = n$ and $|E| = |E(G)| = m$. Let $u, v \in V$, then the edge between these vertices in a simple graph is denoted as $\{u, v\}$, which in shorthand form is often written as uv or $u - v$. In the case of simple graphs, we consider unordered pairs, i.e. $uv = vu$ or $u - v = v - u$. Note that the vertices

$u, v \in V$ are endpoints of edges uv, they are adjacent to each other and simultaneously incident to the edge they form. However, if the pairs of vertices are ordered, i.e. $uv \neq vu$, then we speak of directed graphs (or digraphs). In simple graphs, it is assumed that there can be no more than one edge connecting the same pair of vertices, nor can there be an edge connecting a vertex to itself (a so-called loop). Such structures are referred to as *multigraphs* and *pseudographs*, respectively.

In turn, if we can define weighting functions on edges or vertices, then we speak of *edge-weighted* and *vertex-weighted* graphs, respectively. In such a case, a graph (G, w_v) with a weight function $w_v : V(G) \to R > 0$ defined for its vertices is called a vertex-weighted graph. On the other hand, an edge-weighted graph (G, w_e) is a graph G with a weight function $w_e : E(G) \to R > 0$ defined for its edges. These definitions will be extended later to weighting functions described not by natural numbers, but by functions of several variables with specific properties.

An important construction in graph theory is a *subgraph of a graph* $G(V, E)$. This is a graph $H = (W, F)$ such that $W \subseteq V$ and $F \subseteq E$. In other words, all vertices of graph H belong to $V(G)$, and its edges belong to $E(G)$. Sometimes we can obtain a subgraph of a given graph by removing some edges or vertices.

If we take a fixed set of vertices, say $A \subseteq V$ of a graph G, then if we remove a certain number of vertices from G and all edges coming from and going to them, we obtain an induced subgraph $H = G[A]$. That is, a particular subset of the vertices of G induces such a subgraph. Similarly, if we specify a fixed set of edges, say $F \subseteq E$ of the graph G, then when we remove edges belonging to the set F, a subgraph induced by the edges $H = G - F$ is created. In other words, a subgraph induced by a set of edges in the graph G is understood as a graph whose set of edges is contained (is a subset) in the set of edges of the graph G, and the set of vertices is the endpoints of these edges.

The specific construction is *line graph* $L(G)$ of graph G. This graph is built according the assumptions: each vertex of $L(G)$ represents an edge of G, and two vertices of $L(G)$ are adjacent if and only if their corresponding edges are incident in G.

An important aspect in modelling network structures is the use of relations between vertices and edges. The set of all vertices of the graph $G(V, E)$ adjacent to vertex $v \in V(G)$ is called *a neighbourhood* and denoted $N_G(v)$ or $N(v)$. For any subset $A \subseteq V$ of the set of vertices in graph G, the neighbourhood is defined as $N_G(A) = \bigcup_{v \in A} N_G(v)$. In turn, the set $N_G(v) \cup \{v\}$ is called *a closed neighbourhood* of vertex $v \in V$ in graph G and denoted $N_G[v]$. One of the basic parameters for a graph is the *degree of a vertex and the graph*. The first term denotes the number of vertices belonging to the set $N_G(v)$, and we denote this quantity as $\deg_G(v) = |N_G(v)|$.

In the case of the degree of a graph, we distinguish between the concepts of the smallest and greatest degree of a graph G. The smallest degree of a graph G is given by the number $\delta(G) = \min\{\deg_G(v) | v \in V\}$, and the greatest degree of a graph G is given by the number $\Delta(G) = \max\{\deg_G(v) | v \in V\}$.

Furthermore, for a vertex $v \in V$, the set of all edges incident to it is defined by $I_G(v)$. In other words, this is the set of all edges for which vertex v is one of the endpoints.

The main tasks of CNIs are to ensure the transmission of goods, people, energy, and information between their nodes, executing journeys/information or energy transfer between selected points in this network. Therefore, a crucial issue in modelling these networks using graphs is describing the methods of moving from one node to another. Thus, definitions of routes, paths, and cycles, as well as specific other additional characteristics related to these concepts, must be introduced.

A *route* in a given graph G is a finite sequence of edges of the form $v_o v_1$, $v_1 v_2, \ldots,$ $v_{m-1} v_m$ or $v_0 \rightarrow v_1 \rightarrow v_2 \rightarrow \ldots \rightarrow v_m$, in which every two consecutive edges are either adjacent or identical. This gives us a sequence of vertices $v_0, v_1, v_2, \ldots, v_m$, where vertex v_0 is the starting vertex and v_m is the ending vertex of the route. The number of edges in a route is its length.

Usually, the concept of a route is too broad or general, so the concepts of a *path* and a *road* are defined. A route whose edges are all distinct is a path. A road, on the other hand, is a path whose vertices $v_0, v_1, v_2, \ldots, v_m$ are distinct.

A graph $P_n = (V, E)$ is called a simple path of length n if and only if there is such a numbering of vertices $V = \{v_0, v_1, v_2, \ldots, v_n\}$ that $E = v_o v_1$, $v_1 v_2, \ldots$ $v_{n-1} v_n$.

We can also say that a path or road is closed if $v_0 = v_m$ – then we are dealing with a cycle. Formally speaking, a graph $C_n = (V, E)$ is called a simple cycle of length n if and only if there exists a vertex numbering $V = \{v_0, v_1, v_2, \ldots, v_{n-1}\}$ such that $E = v_o v_1$, $v_1 v_2, \ldots$, $v_{n-1} v_0$.

In the modelling and subsequent analysis of network structures, a crucial concept is *graph connectivity*. A graph G is called connected if and only if a path connects every pair of its vertices.

In addition to the degree of a vertex, structural attributes also include the *number of cycles* and the *diameter of the graph*. The first of these is the evident multiplicity. The diameter of a connected graph G, denoted $d(G)$, is the distance between the two most distant vertices of the graph. In the case of simple graphs, this distance is expressed as the number of edges between these two vertices. In other words, any two vertices of the graph are connected by a path no longer than the diameter.

A well-known approach in the literature to studying connected graphs is to examine their degree of connectivity. It allows us to ask how many edges or vertices must be removed from a graph to make it disconnected. This means that the graph is then partitioned into its connected components. Formally, the connected component of a graph G is the largest connected subgraph of that graph, and each vertex and each edge belongs to exactly one connected component. Therefore, it is true to say that every connected graph has exactly one connected component.

Moreover, the following questions help answer the previously asked question about how many edges or vertices are required to cause at least two components of connectivity. A *cut-set* of a connected graph G is the set of edges whose removal causes G to cease to be connected. A cut in a graph G is a disconnecting set such that no proper set is a disconnecting set. If a cut consists of a single edge, this edge is called a bridge. If G is assumed to be connected, then its edge connectivity $\lambda(G)$ is the number of edges belonging to the smallest cut of G. Furthermore, we say that G is k-edge-connected if $\lambda(G) \geq k$, $k \in N$.

The definitions of the above concepts can be easily extended to graphs with at least two connectivity components, and analogous concepts can be easily obtained for vertex removal.

A *separating set* of a connected graph G is a set of vertices whose removal causes the graph G to cease to be connected. If a separating set consists of exactly one vertex, then that vertex is a separating vertex. It should be clarified that removing a vertex also involves removing all edges entering or leaving it.

Let a graph G be connected, then its vertex connectivity $\kappa(G)$ is the number of vertices belonging to the smallest separating set of G. We say that a graph G is k-vertex-connected or k-connected if $\kappa(G) \geq k$, $k \in N$.

In the case of digraphs, the definitions relating to connectivity require further clarification because they concern ordered pairs of vertices. They are as follows:

1. A directed graph G is *weakly connected* if, after replacing all its directed edges with undirected edges, we obtain an undirected connected graph.
2. A directed graph G is *connected* if it contains a directed path from vertex u to vertex v or a directed path from vertex v to vertex u for each pair of vertices u and v belonging to the set V.
3. A directed graph G is *strongly connected* if it contains a directed path from vertex u to vertex v and a directed path from vertex v to vertex u for each pair of vertices u and v belonging to the set V.

We call a component of the connectedness of a directed graph G is a strong component if it is the largest strongly connected subgraph. It should be noted that $\kappa(G) \leq \lambda(G)$ and both types of connectivity are less than or equal to the lowest degree of the graph G.

Certain types and kinds of graphs are useful in modelling CNIs, the definitions of which are given below.

We start from a planar graph (for example, graphs shown in Figure 2.1c, d, and e). It is a graph in which the only edge intersections occur at the vertices of the graph.

Next, a simple graph in which an edge connects every pair of distinct vertices is a *complete graph* (see 2.1b. The symbol K_n describes a complete graph with n vertices.

An *r-regular graph* is a simple graph in which every vertex has the same degree r, i.e., $\underset{v \in V}{\forall} \deg_G(v) = r$.

A *cyclic graph* is a connected, 2-regular graph. The symbol C_n denotes a cyclic graph with n vertices. Furthermore, a tree is a connected acyclic graph. The symbol T_n means a tree with n vertices.

Infinite graphs are also used to represent the structure of CNI, particularly transportation networks. The most well-known representatives of this family of graphs are *lattice graphs (mesh graphs)*. Lattice graphs are r-regular graphs embedded in Euclidean space, creating their division into regular shapes (quadrangles, pentagons, hexagons, etc.). An essential feature of this family of graphs is the fact that they are regular, i.e., each edge can be assigned the same weight or represent the same distance in Euclidean space or in other spaces.

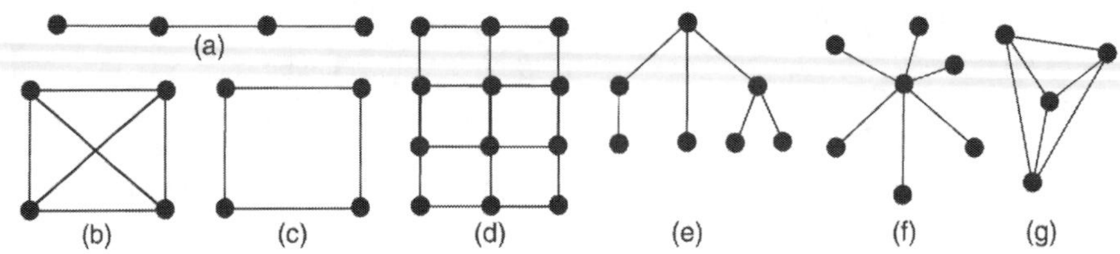

Figure 2.1 Examples of the graphs families

These families of graphs refer to the most common types of real-world CNI topology (see Figure 1.2).

Mesh networks (represented by mesh graphs) – networks in which there are at least two nodes with two or more connections between them (2.1d), which model the road network of most cities.
Hub-and-spoke networks or star networks (represented by star; central node connected to other nodes (2.1f) – networks in which peripheral nodes are connected exclusively to a central node, e.g., computer star network.
Linear networks (represented by paths) – networks in which there is precisely one arc between pairs of consecutive nodes, and each node has exactly two degrees (Fig. 2.1a), which model linear ocean container connections, bus networks.
Tree networks (represented by tree graphs) – nodes are arranged hierarchically, from the highest single node called the root (2.1e, f), which is an example of a network model with at least two hub nodes.
Ring networks (represented by cyclic graph) – nodes constitute close path (cycle).

When we use an edge-weighted graph to model a CNI, we can also find a *minimum spanning tree (MST)*. Generally, a *spanning tree* of a connected graph G is a connected, acyclic subgraph of G that spans every vertex. In the case of an edge-weighted graph, an MST is a spanning tree whose weight (the sum of the weights of its edges) is no greater than the weight of any other spanning tree. We can find it with well-known algorithms: Kruskal's and Prim's (see Algorithms 2.1 and 2.2.

Algorithm 2.1: Kruskal's Algorithm

1. Find the cheapest edge in the graph (if there is more than one, pick one at random). Mark it with any given colour, say red.
2. Find the cheapest unmarked (uncoloured) edge in the graph that doesn't close a coloured or red circuit. Mark this edge red.
3. Repeat Step 2 until you reach out to every vertex of the graph (or you have n-1 coloured edges).
4. The red edges form the desired minimum spanning tree.

Algorithm 2.2: Prim's Algortihm

1. Pick any vertex as a starting vertex – v_{start} . Mark it with any given colour (red).
2. Find the nearest neighbour of v_{start} (call it P1). Mark both P1 and the edge v_{start} P1 red. Cheapest unmarked (uncoloured) edge in the graph that doesn't close a coloured circuit. Mark this edge with same colour of Step 2.
3. Find the nearest uncoloured neighbour to the red subgraph (i.e., the closest vertex to any red vertex). Mark it and the edge connecting the vertex to the red subgraph in red.
4. Repeat Step 3 until all vertices are marked red.
5. The red subgraph is a minimum spanning tree.

Furthermore, CNI modelling using directed graphs allows for finding the maximum flow in the modelled network. This is achieved using the Ford-Fulkerson algorithm. To use it, some additional concepts and formalisms need to be introduced.

Let $G_{st} = (V, A, s, t)$ be a network with $s, t \in A$ being the source and the sink of G_{st}, respectively. Then the capacity of an edge of network G_{st} is mapping $c : A \rightarrow R^+$, which represents the maximum amount of flow that can pass through an edge. It is denoted as c_{uv}, where $u, v \in A$.

A flow in network G_{st} is mapping $f_{uv} : A \rightarrow R^+$, where $u, v \in A$, which subjects to the following two constrains:

- $\underset{(u,v)\in A}{\forall}\ f_{uv} \leq c_{uv}$, (capacity constraint: the flow of an edge cannot exceed its capacity)
- $\underset{v\in V\setminus\{s,t\}}{\forall}\ \sum_{u:(u,v)\in A} f_{uv} = \sum_{u:(u,v)\in A} f_{vu}$, (conservation of flows: the sum of the flows entering a node must equal the sum of the flows exiting a node, except for the source and the sink nodes).

Finally, the value of flow $|f_{uv}| = \sum_{v:(s,v)\in A} f_{sv}$, where s is the source of G_{st}.

Generally, the main problem in traffic flows is maximise value of $|f_{uv}|$, which is called *maximum flow problem*. One of the solutions is using residual network, which we define as follows.

For given G_{st} and flow f_{uv}, we define the *residual network* G_{st} according the following steps:

1. The node set of G_{st}^f is the same as that of G_{st}.
2. Each edge $e = (u, v)$ of G_{st}^f is with capacity $c_e - f_e$.
3. Each edge $e' = (v, u)$ of G_{st}^f is with capacity f_e.

Furthermore, $c_p^f = \min\{c_{uv}^f : (u, v) \in p\}$ is residual capacity of path p in network G_{st}^f.

According to above definitions, the algorithm of finding the maximal flow in network is given in Algorithm 2.3.

Algorithm 2.3: Ford-Fulkerson Algorithm

1. For each edge $e = (u, v)$do
2. Begin
   a. f_{uv}=0
   b. f_{vu}=0
3. End
4. While exists path p from s do t in G_{st}^{f} do
5. Begin
   a. $c_p^f = \min\{c_{uv}^f : (u, v) \in p\}$
   b. For each $e = (u, v) \in p$ do
   c. begin
      i. f_{uv}=f_{uv}+c_p^f
      ii. f_{vu}=-f_{uv}
   d. End
6. End
7. Return f.

2.1.2 GAME THEORY BASIC NOTATIONS

Game theory, in general, is the study of strategic decision-making and action in both conflict and cooperation. This theory provides a good overview of the issues involved in modelling crisis situations for the analysis of critical infrastructures in general.

Due to their specific properties and characteristics, we distinguish many types of games. There are also various ways to classify them. Below are examples of game types and their classifications based on their various characteristics. Of course, the following is informative and by no means exhaustive. Therefore, games can be divided into:

By decision-making time (order):

Strategic (normal) form games – these describe situations in which players make decisions simultaneously, without knowledge of the decisions of other participants.

Extensive (extended) form games – these describe situations in which players make decisions sequentially, at successive moments in time, having specific information about the decisions of other players (and their own) at previous moments.

Due to the knowledge available:

Games with complete information – these are games in which players have full information about the possible outcomes of the game (they know their own and other players' payoff functions) and about the sets of possible player strategies. In the case of extensive games, where players have complete information at all times about other players' previous decisions and their potential random moves, we refer to these as games with complete information.

Games with incomplete information are the opposite of those indicated above.

Due to the possibility of forming coalitions

Cooperative (coalition) games – in this case, actions are assigned to groups (coalitions) of players.

Non-cooperative games – in this case, actions are assigned to individual players.

Based on the number of players, we distinguish *two-player* games and *infinite multiplayer* games (in particular, so-called "large games," i.e., games with a continuum of players).

Based on the sets of available actions and strategies:

Finite games – when each player's set of actions or strategies is finite.

Infinite games – in particular, we distinguish games with a continuum of actions (strategies).

Based on the number of executable actions: Games with a *finite* and *infinite* time horizon.

Based on recurrence: *Single iteration* and *multiple (iterated) games.*

Based on the "role" of time: *static* and *evolutionary* games.

Other game characteristics include stochastic games, differential games, dynamic games, games against nature, and games on networks.

Let's start form some basics. The strategic game consists of three elements: a set of players, a set of actions (strategies) for the players, and payoffs. Therefore, when building a model based on game theory, we must propose a set of players, a set of strategies and a matrix of payoffs associated with individual strategies. More formally, we introduce a set of players – $PL = \{1, 2, \ldots, n\}$, a non-empty set of player i actions (pure strategies) – A_i, $A = \times A_i$, and player i payoff function – $u_i : A \rightarrow \Re$, $i = 1, 2, \ldots, n$. Additionally, we describe a set of a game's pure strategies profile as $a = (a_1, a_2, \ldots, a_n) = (a_i)_{i \in PL}$, and payoff for player i based on profile a as $u_i(a)$.

Sometimes, to distinguish player i, e.g., to compare the payoff function values in profiles where we change one coordinate, we will write the profile as $(a_i, _{-i})$, where a_{-i} denotes the sequence of profile terms (a_j) for all players except$i$: $a_{-i} = (a_j)_{j \in N \setminus i}$. Consequently, we denote$A_{-i} = \times A_k$, $k \in N \setminus i$.

Taking the above into account, we can define the Nash equilibrium in pure strategies. This is an action profile for pure strategies $a^* = (a_1^*, a_2^*, \ldots, a_N^*) \in A$ such that

$$\forall_{a_i \in A_i} u_i(a_i^*, a_{-i}^*) \geq u_i(a_i, a_{-i}^*) \tag{2.1}$$

In other words, no player can increase his payoff by unilaterally – that is, without changing the strategies of all other players – changing their strategy.

However, using pure strategies is not always possible. They often do not correspond to the described reality. In such cases, a mixed strategy should be employed. Formally, the mixed strategy σ_i of player i in the strategic game is the probability distribution on the set of his pure strategies A_i, i.e.,

$$\sigma_i = (\sigma_{i\,1}, \sigma_{i\,2}, \ldots, \sigma_{i\,m_i}). \tag{2.2}$$

The coordinate $\sigma_{ih} \geq 0$ is the probability that player i will play a pure strategy (choose action) $h \in A_i$. We introduce the following notations:

$\Sigma_i = \{\sigma_i : A_i \to [0,1] : \Sigma_{k=1}^{m_i} \sigma_{ik} = 1, \sigma_{ik} \geq 0\}$ – set of mixed strategies of player i
$\sigma = (\sigma_j)_{j \in N} = (\sigma_1, \sigma_2, \ldots, \sigma_n)$ - game profile
$\Sigma = \times \Sigma_i$, $i \in N$ – set of all game profiles
$\sigma_{-i} = (\sigma_1, \sigma_2, \ldots, \breve{\sigma}_i, \ldots, \sigma_N)$ – strategy profile of all players except player i
$u_i(\sigma) = u_i(\sigma_i, \sigma_{-i})$ - player i's payoff from profile σ.

Now assume that $a = (a_1, a_2, \ldots, a_n)$, $a_i \in A_i$ is the pure strategy profile of the strategic game. According to statistical independence, the (joint) probability that player i will choose action a_1, player 2 will choose action a_2, etc., is given by

$$x(a) = x_{1a_1} \cdot x_{2a_2} \cdots x_{na_n} \tag{2.3}$$

where

x_{ia_i} is the probability that player i will play strategy a_i, $i = 1, \ldots, n$
$x(a) \in [0,1]$
$\Sigma_{a \in A} x(a) = 1$.

Note: Each player plays their strategy independently. Hence, the assumption of statistical independence in Eq. 2.3.

Thus, for each player i on their set of pure strategy profiles, we define a random variable U_i with the following distribution:

$$(u_i(a), x(a)),\ a \in A \tag{2.4}$$

where

$u_i(a)$ – player i's payoff from strategy a item $x(a)$ – probability of playing profile a, according to the formula 2.3

At the same time, the payoff of player i from the mixed strategy profiles $x = (x_1, \ldots, x_n)$ is expressed as the expected value of the random variable U_i:

$$\bar{u}_i = E[U_I] = \sum_{a \in A} x(a) u_i(a) \tag{2.5}$$

Table 2.1 shows the loss to Player A from Strategy 2 (the lower row) – based on Player B then playing Strategy II (the right column). This is a game in normal form,

Table 2.1
Normal form Payoff Matrix for a Single-Stage Game Two-Target Game

		Player B	
	Strategies	**Strategy I**	**Strategy II**
Player A	Strategy 1	10	-20
	Strategy 2	30	20

with Player A choosing a row and Player B choosing a column. The sets of rows and columns represent the two players' respective pure strategy sets.

In Table 2.2, in each cell, the first number is the expected utility or "payoff" to the Player A or row-choosing player, and the second number is the expected utility or "payoff" to the Player B or column-choosing player. When Player A chooses Strategy 2 and Player B Strategy II, then Player A first gets a payoff of $u_{22}(A)$ and Player B gets $u_{22}(B)$.

Table 2.2
Payoff Matrix for a General 2×2 Bi-matrix Game

		Player B	
	Strategies	**Strategy I**	**Strategy II**
Player A	Strategy 1	$u_{11}(A), u_{11}(B)$	$u_{12}(A), u_{12}(B)$
	Strategy 2	$u_{21}(A), u_{21}(B)$	$u_{22}(A), u_{22}(B)$

Note: $u_{ij}(A)$ – payoff of Player A, $u_{ij}(B)$ – payoff of Player B, $i, j = 1,2$

In strategic games, we define strictly dominant and weakly dominant strategies. Using the notations introduced earlier, we have the following definitions. Strategy $\sigma_i \in \sum_i$ strictly dominates strategy $\eta_i \in \sum_i$ if

$$\forall_{\sigma_{-i} \in \sum_{-i}} u_i(\sigma_i, \sigma_{-i}) > u_i(\eta_i, \sigma_{-i}) \tag{2.6}$$

Strategy $\sigma_i \in \sum_i$ weakly dominates strategy $\eta_i \in \sum_i$ if

$$\forall_{\sigma_{-i} \in \sum_{-i}} u_i(\sigma_i, \sigma_{-i}) \geq u_i(\eta_i, \sigma_{-i}) \tag{2.7}$$

We say that the corresponding strategies η_i are strictly (weakly) dominated by the above strategies σ_i. A strategy is weakly dominated if there is another strategy that weakly dominates it.

Introducing the above concepts provides us with a simple algorithm for solving games, as shown in Tables 2.1 and 2.2. The steps of this algorithm are as follows:

1. If there is no player with a strictly dominated strategy, stop. Otherwise, go to point 2.
2. Remove this strategy and return to point 1.

Here, we have to make one remark. In case of the game given in Table 2.1, in row perspective, strictly (weakly) dominated strategies are defined as above. But in column perspective, inequalities are opposite, because the payoff (≥ 0) is the win of Player A, and it is the loss of Player B (we minimise the lost).

Now, we go to focus on making multi-criteria decisions in conditions of uncertainty – it is a challenge for decision-makers. For this purpose, you can use a *game with nature* tool in the literature (Simon, 1959; Papadimitriou, 1985; Raiffa, 1968; McNamee and Celona, 2008; Pazek and Rozman, 2009). It is a two-player game in which one of the players is not interested in the outcome and, simultaneously, does not take risks that may occur with a certain probability.

The player called nature is indifferent to what results both they and their adversary receive, but he does not take any risk when entering the game. However, the player-decision-maker must consider the so-called states in nature, thus operating under uncertain conditions. The literature defines a state by nature as phenomena that occur regardless of the player's expectations and often manifest themselves after choice or implementation. Conditions inherently determine the outcome but are not controlled by the player concerned during the examination.

To solve the game with nature, we use commonly known criteria (Simon, 1959; Papadimitriou, 1985; Raiffa, 1968; McNamee and Celona, 2008; Pazek and Rozman, 2009):

Wald's
optimistic
Hurwicz's
Savage's
Laplace-Bayes's

The decision-maker in *Wald's criterion assumes* that a worst-case scenario (state of nature) will occur, leading to the worst possible payoff for a given player's strategy. The strategy that maximises this minimum payoff is then chosen. This is a maxmin strategy. It characterises a player interested in minimising risk and protecting himself against the worst possible outcome in the event of an unfavourable state of nature. The Wald criterion is derived from the following formula (Wald, 1949):

$$v = \min_{i} \left\{ \min_{j}(w_{ij}) \right\} \tag{2.8}$$

where w_{ij} – payoff value for strategy a_i, $i = 1,2,\ldots,n$ (number of rows), $j = 1,2,\ldots,m$ (number of nature's states).

The optimism criterion, also called the *maximax criterion*, represents an approach to decision-making in which the decision-maker aims to maximise the best possible outcome for each state of nature. In the case of the optimism criterion, the decision-maker assumes that the most favourable state of nature (the best possible scenario) will occur and chooses the strategy that brings the highest possible outcome in this case. In other words, the decision-maker is optimistic about future conditions and aims to maximise their potential benefits. This criterion assesses the achieved maximum outcome for each state of nature. This criterion is derived from the following formula (McNamee and Celona, 2008):

$$v = \min_j \left\{ \max_i (w_{ij}) \right\} \tag{2.9}$$

where w_{ij} – payoff value for strategy a_i, $i = 1,2,\ldots,n$ (number of rows), $j = 1,2,\ldots,m$.

The *Hurwicz criterion* (Hurwicz, 1951) is used when the decision-maker must consider both risk (uncertainty) and the possibility of achieving different outcomes. The following formula is the definition of the Hurwicz criterion:

$$v_i = \lambda \min_j \{w_{ij}\} + (1-\lambda) \max_j (w_{ij}) \tag{2.10}$$

where λ is a caution factor, w_{ij} – payoff value for strategy a_i, $i = 1,2,\ldots,n$ (number of rows), $j = 1,2,\ldots,m$.

In practice, the decision-maker selects a strategy that considers the possibility of achieving high results in the best case (optimistic scenario) and minimising risk in the worst case (pessimistic scenario). The Hurwicz criterion allows the decision-maker to consider risk and his preferences regarding expected results when making a decision.

The *Savage criterion* (Savage, 1951), also known as the minimax regret criterion, is another decision-making criterion. This approach minimises the so-called "regret (r)," the difference between the best possible and actual results in any state of nature. The procedure is based on the decision-maker's pessimistic approach, assuming the worst possible scenario. In this approach, the regret value is calculated for each strategy, expressed as the difference between the highest payoff in a given state of nature and the remaining values for that state. The following formula describes this situation:

$$r_{ij} = \max_i (w_{ij}) - w_{ij} \tag{2.11}$$

where w_{ij} is the payoff value for strategy a_i, $i = 1,2,\ldots,n$ (number of rows), $j = 1,2,\ldots,m$.

Then, the highest regret value is selected for each strategy, and in the next step, the lowest of these values is selected, using the formula:

$$v = \min_i \left\{ \max_j (r_{ij}) \right\} \tag{2.12}$$

where $i = 1,2,\ldots,n$ (number of rows), $j = 1,2,\ldots,m$.

Thus, the final strategy is to minimise the expected value of "regret" in the most pessimistic state of nature.

The *Laplace-Bayes criterion*, also known as the maximum expected value criterion, is another decision-making criterion. It is used when the decision-maker has incomplete or imprecise information about the probabilities of various states of nature. According to Laplace-Bayes' criterion, for each state of nature (S_j in S), the decision-maker should evaluate the probability of p_j that S_j will occur. The decision rule is in Algorithm 2.4.

Algorithm 2.4: Laplace-Bayes decision rule

1. Input: game with nature (state of nature $(S_{j)}$, m – number of natue's states, and payoff value);
2. Output: the maximum expected value criterion;
3. $p_j = P(S_j) = 1/m$ to each $S_j \in S$, for $j = 1, 2, \ldots, m$
4. For each payoff matrix row a_i, calculate its expected value: $E(a_i) = \sum_j p_j(w_{ij})$, where w_{ij} – payoff of i-th row and j-th column.
5. To $i = 1, 2, \ldots, n$, because p_j is a constant, $E(a_i) = p_j \sum_j (w_{ij})$
6. Choosing the higher $E(a_i)$ value is the optimal decision under uncertainty.

The Laplace-Bayes criterion is based on treating all possible states of nature equally and is used when more detailed information about probabilities is lacking. It is a simple and intuitive method that allows the decision-maker to choose the strategy with the highest average outcome value, regardless of the probability of the individual states of nature occurring, i.e., according to the formula:

$$v = \max_i [E(a_i)] \tag{2.13}$$

2.2 CRITICAL NETWORK INFRASTRUCTURE MODELLING

The basis of a well-constructed mathematical model of a real-world object is its consistency with the reality being described. In the case of CNIs, as emphasised several times earlier, the best representation of their actual structure is a connected graph (or digraph) in which vertices describe nodes (or junctions) and edges their arcs (or links). When we apply a weighted graphs, we build a functional model. In this way, a weights for vertices and edges represent a functional strains (workload, time of operation, etc). The level of accuracy of these models can vary depending on the problem to be solved or the time available to build the model.

Let's consider a CNI model that focuses on only the vertices corresponding to the most important nodes, and the arcs/links drawn between them represent paths between these nodes. If there is more than one path between these nodes, parallel edges are created. Thus, a graph constructed in this way is a multigraph, as discussed earlier in this chapter. It can be simple of directed graph. An example of such a network model is shown in Figure 2.2, where P_{il_i} denotes the i-th path between selected nodes.

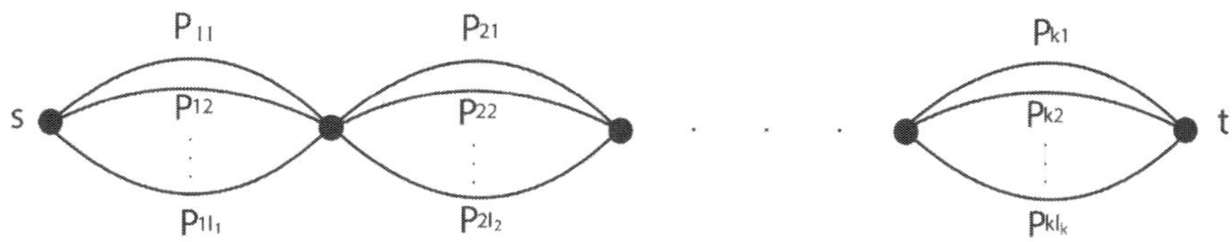

Figure 2.2 A model of a network described by a multigraph with possible paths between selected vertices

The CNIs model shown in Figure 2.2 is the *parallel-series* type. This means that its vertices form a serial subsystem, while its edges form a parallel structure. For such a network, a single path P_{il_i} consists of l_i vertices and $i \in N$. This gives us a model of the CNI in which important possible connections between its source and sink exist. This model is shown in Figure 2.3.

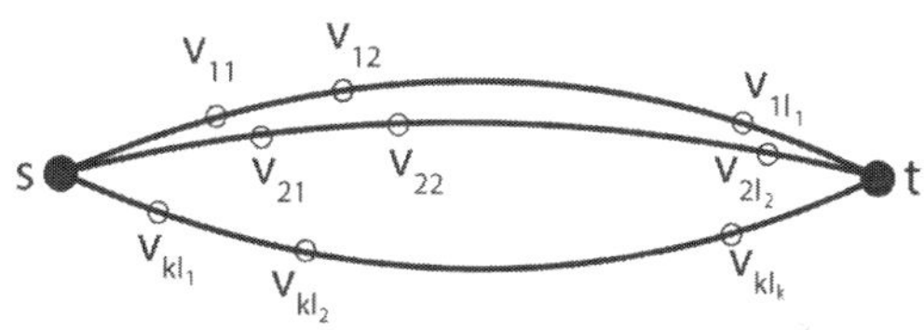

Figure 2.3 A transport network model that takes into account the connections between the source and the sink, without distinguishing intermediate nodes

As can be easily seen, the model of the connection between the network source and sink is a series-parallel model in the literature on reliability modelling, including (Kołowrocki and Soszyńska-Budny, 2011). Its vertices: v_{il_i}, $i \in N$, on the paths form subsystems with a serial structure, and the entire paths are parallel to each other.

When we are interested in functional modeling of CNIs, the best approach is to use weighted graphs. The definitions of weights in graphs presented above (see Section 2.1) enable the representation of the load on individual network elements using natural numbers. These can represent maximum capacity, average travel time, connection importance, travel cost, reloading rate, and other characteristics.

Modelling CNI using weighted graphs in the traditional way of understanding can be inadequate, particularly when addressing analysis or optimisation problems, especially introduced in Chapter 4. In some instances, the typical method of assigning edge or vertex weights may introduce errors into the results or be overly simplistic. To address these issues, we define a new family of weighted graphs with two subtypes, based on the concept of continuous weight functions for vertices and edges.

A real-valued function of n-variables, continuous, differentiable at least class C^2, and integrable, $W_V : D_{W_V} \to \mathbb{R}$, $D_{W_V} \subseteq \mathbb{R}^n$, is called the *continuous vertex-weighted function (CVWF)* of the graph $G(V,E)$.

A real-valued function of n-variables, continuous, differentiable at least class C^2, and integrable, $W_E : D_{W_E} \to \mathbb{R}$, $D_{W_E} \subseteq \mathbb{R}^n$, is called the *continuous edge-weighted function (CEWF)* of the graph $G(V,E)$.

The family of graphs with vertex weights as described above is termed *continuous-vertex-weighted graphs* (CVWG). Similarly, the family of graphs whose edge weights adhere to continuous edge-weighted function is named *continuous-edge-weighted graphs* (CEWG).

These two graph families broaden the scope for describing challenges encountered in CNIs. The proposed approach applies to any system represented by connected, simple, and weighted (both directed and undirected) graphs.

However, the assumptions outlined in above definitions regarding the properties of weight functions do not ensure that they will definitively solve the modelled problem. The benefits of such a description become apparent in addressing optimisation problems within CNIs. Although graphs are discrete combinatorial structures traditionally analysed using discrete optimisation methods, the newly defined graph families facilitate the application of established linear and nonlinear optimisation techniques.

2.3 MODELLING RELIABILITY, VULNERABILITY, AND SAFETY OF CRITICAL NETWORK INFRASTRUCTURE

At the very beginning of this section, we introduce the topological reliability of the CNIs.

The critical network infrastructure is *reliable in topological way*, allowing for the transmission of goods, information, or energy between any nodes, i.e. if travel or sending data/energy is possible from any node in the network to any other.

The approach to the reliability of CNIs, presented later in this book, is based on connection reliability (Bell and Iida, 1997; Iida, 1999; Berdica, 2002; Chen et al., 2002; Hongwei and Xizhao, 2015). However, the advantage of the presented approach and the definitions formulated below is the ability to take into account the multi-state approach when analysing the reliability structures of CNIs. This approach, however, is preceded by the introduction of topological reliability based on solutions related to the structural analysis of CNI. It is expressed using a structural function, described in two variants: vertex and edge.

For the notions proposed later in this book, we assume that a reliable CNI is a connected network (with one-component connectivity), meaning that there is a spanning tree in this network. In other words, it is represented by a connected graph.

Consider a CNI represented by an undirected graph $G(V,E)$, where $|V| = n$, $|E| = m$. We consider the reliability structures discussed in Section 1.4:

serial
parallel
"*k* out of *n*"
"consecutive *k* out of *n*: F".

Each of the structures mentioned will have two definition variants:

edge – when the network behaviour is considered based on edges;
vertex – when the network behaviour is considered based on vertices.

The division into variants allowed the development of simple algorithms for identifying CNIs reliability structures, which are proposed later in Section 2.2.

Assuming the connectivity of the graph representing the CNI, the vertex variant of the definition of CNI structures defines it as follows.

We say that a CNI is:

- a vertex-serial if there is precisely one vertex whose removal will result in the creation of at least two connectivity components (a so-called separating vertex) in the graph G.
- a vertex-parallel if the creation of at least two connectivity components requires the removal of $n-1$ vertices in the graph G.
- a vertex-"k out of n " if the removal of any k of the n available vertices will result in the creation of at least two connectivity components in the graph representing it.
- a vertex-"consecutive k out of n: F" if the removal consecutive k of any of the n available vertices will result in the creation of at least two connectivity components in the graph representing it.

The operation of deleting a vertex, as indicated above, means removing it along with all edges incident to it. Additionally, the phrase "consecutive vertices" refers to vertices that are adjacent to each other (i.e., there is an edge between them).

To understand the above concepts, let's use the graphs shown in Figure 2.1. In the case of a path 2.1A, a tree 2.1E, or a star 2.1F, removing a single vertex is enough to create two connectivity components in these graphs, so they have a vertex-serial structure. However, if we consider the complete graph in Figure 2.1b and the wheel 2.1G, removing individual vertices does not create two connectivity components. Therefore, they have a vertex-parallel structure. In turn, the cycle in Figure 2.1c requires the removal of any two, but not consecutive, vertices, i.e., it has a vertex-"2 out of 4" structure. The lattice graph in Figure 2.1d requires the removal of at least three consecutive vertices (one of the internal horizontal lines of vertices; they cannot be external) to create two connectivity components. Therefore, it has a vertex-"consecutive 3 out of 12:F" structure.

And now if we generalise the above results, we can formulate Theorem 2.1 as follows:

Theorem 2.1 *A critical network infrastructure represented by*

1. a complete graph K_n is the vertex-parallel network.

2. a cyclic graph C_n is vertex-"2 of n," where n denotes the number of vertices.

3. a tree T_n, a path P_n, or star S_n is a vertex-serial network.

4. a wheel W_n is a vertex-"consecutive 3 out of n:F."

Note:

Ad 1. The minimum number of vertices necessary to generate at least two connectivity components is $n-1$. Hence, the above statement is true.

Ad 2. The minimum number of vertices necessary to generate at least two connectivity components is 2. Hence, the above statement is true.

Ad 3. The minimum number of vertices necessary to generate at least two connectivity components is 1 – meaning there is always a bridge in the graph. Hence, the above statement is true.

Ad 4. The minimum number of vertices necessary to generate at least two connectivity components is 3 in the series. This must necessarily include a central vertex. Hence, the above statement is true.

In the case of the edge variant, the definitions of the reliability structures of CNI are as follows.

A CNI is:

- an edge-serial if there is exactly one edge whose removal will result in the creation of at least two connected components (a so-called bridge) in the graph *G*;
- an edge-parallel if the creation of at least two connected components in the graph *G* requires the removal of $m-1$ edges;
- an edge-"*k* out of *m*" type if the removal of any *k* edges from the existing *m* edges will result in the creation of at least two connected components;
- an edge-"consecutive *k* of *m*: F" type if the removal of *k* consecutive adjacent edges from the existing *m* edges will result in the creation of at least two connected components.

Similarly to the vertex variant, the term "consecutive edges" refers to edges that are incident to each other, i.e., those that share a common vertex. In particular, they will form trees, paths, or cycles.

To understand the concepts of reliability of CNIs in edge sense, let's use the graphs in Figure 2.1 one more time. In the case of a path 2.1A, a tree 2.1E, or a star 2.1F, removing a single edge is enough to create two connectivity components in these graphs. So they have a edge-serial structure. However, if we consider the complete graph in Figure 2.1b, removing three edges creates two connectivity components (one isolated vertex). Therefore, they have a edge-"consecutive 3 out of 6:F" structure. In turn, the cycle in Figure 2.1c requires the removal of any two edges, i.e., it has a edge-"2 out of 4" structure. The lattice graph in Figure 2.1d requires the removal of at least two consecutive edges (to isolate one of the vertices) to create two connectivity components. Therefore, it has a edge-"consecutive 2 out of 17:F" structure.

And now if we generalise the above results, we can formulate Theorem 2.2 for CNI topological reliability in edge sense.

Theorem 2.2 *A CNI represented by*

1. a complete graph K_n is an edge-"consecutive $n-1$ of $\frac{n(n-1)}{2}$:F" network.

2. a cyclic graph C_n is an edge-"2 of m" network, where m denotes the number of edges.

3. a tree T_n, a path P_n, or star S_n is an edge-serial network.

Note:

Ad 1. The minimum number of edges necessary to generate at least two connectivity components is $n-1$. Hence, the above statement is true.
Ad 2. The minimum number of edges necessary to generate at least two connectivity components is 2. Hence, the above statement is true.
Ad 3. The minimum number of edges necessary to generate at least two connectivity components is 1 – meaning there is always a bridge in the graph. Hence, the above statement is true.

The next step in formulating a new description of the reliability of CNI is to define the reliability function for the structures of these infrastructures, as described in Section 1.3 To achieve this, we adopt the following notations:

- n is the number of vertices/nodes of the CNI
- m is the number of edges/arcs of the CNI
- V_{i_v}, $i_v = 1, 2, \ldots, n$ are nodes/vertices of the CNI; in the two-state approach, all nodes of the network, as well as the network itself, can be either operational or inoperative; in the multi-state approach, all nodes of the network, as

well as the network itself, can reside in a set of reliability states $\{0,1,\ldots,z\}$, $z \geq 1$, where the states are ordered from state 0 (the worst) to state z (the best); In the mixed approach, network nodes are either operational or non-operational, and the entire transport network can reside in a set of reliability states, as in the multi-state approach

- E_{i_e}, $i_e = 1,2,\ldots,m$ are arcs (edges) of the CNI; in the two-state approach, all network arcs, as well as the network itself, can be operational or non-operational; in the multi-state approach, all network arcs, as well as the network itself, can reside in a set of reliability states $\{0,1,\ldots,z\}$, $z \geq 1$, where states are ordered from state 0 (the worst) to state z (the best); in the mixed approach, network nodes are operational or non-operational, and the entire CNI can reside in a set of reliability states, as in the multi-state approach
- $v_{i_v}^{bin}$ is a binary random variable describing the state of the nodes/vertices of the CNI V_{i_v}, $i_v = 1,2,\ldots,n$
- $e_{i_e}^{bin}$ is a binary random variable describing the state of the edge/arc of CNI E_{i_e}, $i_e = 1,2,\ldots,m$
- $\varphi_s(\cdot)$ is a structural function of the serial CNI
- $\varphi_p(\cdot)$ is a structural function of the parallel CNI
- $\varphi_{k|n}(\cdot)$ is a structural function of the "k out of n" CNI
- $\varphi_{C-k|n:F}(\cdot)$ is a structural function of the "consecutive k out of n: F" CNI.

In order to describe the reliability of such defined structures, one can use the structural function described, among others, by Iida (1999) and Rausand and Høyland (2004) and proposed earlier by Mine and Kawai Iida (1999); Mine and Kawai (1982). Based on this, we assume that:

- the state of vertex v is represented by a random binary variable $v_{i_v}^{bin}$, $i_v = 1,2,\ldots,n,$, i.e.:

$$v_{i_v}^{bin} = \begin{cases} 1 & \text{if vertex } V_{i_v} \text{ is accessible,} \\ 0 & \text{if vertex } V_{i_v} \text{ is not available,} \end{cases} \tag{2.14}$$

- the state of the network edge e is represented by some binary random variable $e_{i_e}^{bin}$, $i_e = 1,2,\ldots,m$, i.e.:

$$e_{i_e}^{bin} = \begin{cases} 1 & \text{if edge } E_{i_e} \text{ is available,} \\ 0 & \text{if edge } E_{i_e} \text{ is not available.} \end{cases} \tag{2.15}$$

Furthermore, we assume that the state of the CNI is described by a structured function $\varphi(\mathbf{v})$ or $\varphi(\mathbf{e})$, where $\mathbf{v} = [V_1, V_2, \ldots, V_n]$ is a vector of random variables representing vertices in the transport network and $\mathbf{e} = [E_1, E_2, \ldots, E_m]$ is a vector of random variables representing links in the transport network. We assume that if the network is functional, the value of the structural function is 1, and if it is inefficient, the value of this function is 0. Therefore, based on Iida (1999), we have:

- for the system state in the vertex sense:

$$\varphi(\mathbf{v}) = \begin{cases} 1 & \text{if network is vertex-functional,} \\ 0 & \text{if network is vertex-inefficient,} \end{cases} \tag{2.16}$$

- for the system state in the edge sense:

$$\varphi(\mathbf{e}) = \begin{cases} 1 & \text{if network is edge-eligible,} \\ 0 & \text{if network is edge-inefficient.} \end{cases} \tag{2.17}$$

Then, for the specific structures defined above, the following structural functions are true, allowing for determining the reliability (Aki and Hirano, 1996; Navarro and Eryilmaz, 2007; Rausand and Høyland, 2004):

- edge-serial network

$$\varphi_s(\mathbf{e}) = e_1^{bin} \cdot e_2^{bin} \cdot \ldots \cdot e_m^{bin} = \prod_{i_e=1}^{m} e_{i_e}^{bin} \tag{2.18}$$

- vertex-serial network

$$\varphi_s(\mathbf{v}) = v_1^{bin} \cdot v_2^{bin} \cdot \ldots \cdot v_n^{bin} = \prod_{i_v=1}^{n} v_{i_v}^{bin} \tag{2.19}$$

- edge-parallel network

$$\varphi_p(\mathbf{e}) = 1 - \prod_{i_e=1}^{m} (1 - e_{i_e}^{bin}) \tag{2.20}$$

- vertex-parallel

$$\varphi_p(\mathbf{v}) = 1 - \prod_{i_v=1}^{n} (1 - v_{i_v}^{bin}) \tag{2.21}$$

- edge-"k out of m" network

$$\varphi_{k|n}(\mathbf{e}) = \begin{cases} 1 & \text{jeżeli } \sum_{i_e=1}^{m} e_{i_e}^{bin} \geq k_e, \\ 0 & \text{jeżeli } \sum_{i_e=1}^{m} e_{i_e}^{bin} < k_e, \end{cases} \tag{2.22}$$

- vertex-"k out of m" network

$$\varphi_{k|n}(\mathbf{v}) = \begin{cases} 1 & \text{jeżeli } \sum_{i_v=1}^{n} e_{i_v}^{bin} \geq k_v, \\ 0 & \text{jeżeli } \sum_{i_v=1}^{n} e_{i_v}^{bin} < k_v, \end{cases} \tag{2.23}$$

- edge-"consecutive k out of m: F" network

$$\varphi_{C-k|m:F}(\mathbf{e}) = \prod_{j=i_e}^{m-k+1} \left\{ 1 - \prod_{i_e=j}^{j+k-1} (1 - e_{i_e}^{bin}) \right\} \tag{2.24}$$

- vertex-“consecutive k out of m: F” network

$$\varphi_{C-k|n:F}(\mathbf{v}) = \prod_{j=i_v}^{n-k+1}\left\{1-\prod_{i_v=j}^{j+k-1}(1-v_{i_v}^{bin})\right\} \tag{2.25}$$

Taking into account the above assumption, based on Iida (1999), the reliability of a connection is represented by the expected value of a random variable representing a given arc or node. To define the reliability of a CNI as the probability that the network functions and performs its tasks, serving travel/data/energy requests, it should be expressed by the expected value of the structural function $\varphi(\mathbf{x})$. Assuming that r_e denotes arc reliability, r_v denotes node reliability, and R denotes system reliability, these quantities are expressed using the following formulas (Iida, 1999):

$$R = E\{\varphi(\mathbf{x})\} \tag{2.26}$$

$$r_e = E\{e_{i_e}^{bin}\} \tag{2.27}$$

$$r_v = E\{v_{i_v}^{bin}\} \tag{2.28}$$

The above-proposed approach to describing the reliability structures of CNIs allows for the use of two-state, multi-state, and mixed methods to model the reliability of these networks, also taking into account lifetime as presented in Section 1.4 based on the results (Khan, 2014; Kołowrocki, 2004, 2013; Kołowrocki and Soszyńska-Budny, 2011).

However, in the latter model, the entire CNI is treated as multi-state, but its elements are two-state. This can also be treated as multi-state elements with a set of reliability states composed of two components: state zero and state z, i.e., $u = 0, 1$.

Regarding the results presented in Section 1.4 regarding the reliability of two-state and multi-state technical systems, and based on the results of research on transport networks included in Guze (2017, 2018), in order to model the reliability of CNI treated as complex technical systems, represented by their lifetimes, the following notations are adopted:

- VT_i – independent random variables describing the lifetimes of two-state nodes (vertices) of the system network V_i, $i = 1, 2, \ldots, n$;
- ET_j – independent random variables describing the lifetimes of two-state arcs of the CNI E_j, $j = 1, 2, \ldots, m$;
- NT – independent random variable describing the uptime of a two-state CNI;
- $VT_i(u)$, $i = 1, 2, \ldots, n$ – independent random variables representing the lifetimes of CNI nodes V_i in the subset of reliability states $\{u, u+1, \ldots, z\}$ when they were in state z at time $t = 0$;
- $ET_j(u)$, $j = 1, 2, \ldots, m$ – independent random variables representing the lifetimes of the CNI arcs in the subset of reliability states $\{u, u+1, \ldots, z\}$ when they were in state z at time $t = 0$;

- $NT(u)$ – an independent random variable representing the system's time in the subset of reliability states $\{u, u+1, \ldots, z\}$, assuming that it was in state z at time $t = 0$ and that without repair, the network condition deteriorates over time t;
- $VR_i(t) = P(VT_i > t)$, $t \in < 0, +\infty)$ – the reliability function of the two-state network nodes V_i, $i = 1, 2, \ldots, n$;
- $VF_i(t) = 1 - VR_i(t) = P(VT_i \leq t)$, $t \in < 0, +\infty)$ – the distribution function (reliability function) of the vertices V_i,$i = 1, 2, \ldots, n$;
- $ER_j(t) = P(ET_j > t)$, $t \in < 0, +\infty)$ – the reliability function of the two-state arc of the CNI E_j, $j = 1, 2, \ldots, m$;
- $EF_j(t) = 1 - ER_j(t) = P(ET_j \leq t)$, $t \in < 0, +\infty)$ – the distribution function (unreliability function) of the CNI arc E_j, $j = 1, 2, \ldots, m$;
- $R(t) = P(T > t)$, $t \in < 0, +\infty)$ – the reliability function of the two-state CNI;
- $F(t) = 1 - R(t) = P(T \leq t)$,$t \in < 0, +\infty)$ – the distribution function (unreliability function) of the CNI;
- $s_{e_j}(t)$ – reliability state of edge E_j at time t, $t \geq 0$;
- $s_{v_i}(t)$ – reliability state of vertex element V_i at time t, $t \geq 0$;
- $s_{NT}(t)$ – reliability state of the CNI at time t, $t \geq 0$.

2.3.1 TWO-STATE RELIABILITY MODELS OF CRITICAL NETWORK INFRASTRUCTURE

Depending on the elements that are currently considered relevant to the reliability of critical infrastructure networks, we have two possible definitions of reliability. Considering the notations presented above and the results for complex technical systems presented in Section 1.4, based on formulas 1.9 – 1.21, in the two-state case, the following definitions of critical infrastructure network reliability are accurate.

A two-state complex technical network is called

- vertex-serial (edge-serial) its lifetimes NT are expressed by the following relationship:

$$NT = \min_{1 \leq i \leq n} \{VT_i\} \tag{2.29}$$

$$\left(NT = \min_{1 \leq j \leq m} \{ET_j\} \right) \tag{2.30}$$

- vertex-parallel (edge-parallel), if its lifetimes NT are expressed by the following dependencies:

$$NT = \max_{1 \leq i \leq n} \{VT_i\} \tag{2.31}$$

$$\left(NT = \max_{1 \leq j \leq m} \{ET_j\} \right) \tag{2.32}$$

- a vertex-"l out of n" (edge-"l out of m"), if its time of availability NT is given as follows:

$$NT = VT_{(n-l_v+1)}, l_v = 1, 2, \ldots, n, (NT = ET_{(m-l_e+1)}, l_e = 1, 2, \ldots, m) \tag{2.33}$$

where $VT_{(n-l_v+1)}(ET_{(n-l_e+1)})$ is the lv-th (le-th) statistic of the highest degree in the sequence of network element lifetimes VT_1, VT_2,..., VT_n (ET_1, ET_2,..., ET_m).

The above means that a system is efficient if its l_v (l_e) vertices (edges) out of n (m) accessible are efficient. In the special case, when m=n, we have a parallel system, and when m=1, we have a serial system.

In the case of the definition of a two-state vertex-"consecutive k out of n: F" (edge-"consecutive k out of m: F") network, it can be expressed using the definition given in Chapter 1 and based on the results obtained by Aki and Hirano (1996), which define this type of structure using the lifetimes.

If we denote by $T_{k|n:F}$ the uptime of the "consecutive k out of n: F" CNI, then the reliability function is expressed as follows (Aki and Hirano, 1996):

$$R_{k|n:F}(t) = P(T_{k|n:F} > t) = 1 - \sum_{i=1}^{n} \omega_i F_{i:n}(t) \tag{2.34}$$

where

$F_{i:n}(t)$ is the cumulative distribution function of the i-th order statistic of the random variable Ti:n, in a random sample $T_1, T_2, \ldots, T_n$ (or the uptime of a threshold transport network of type $(n-i+1)ofn$ with the lifetimes of its data elements as $(T_1, T_2, \ldots, T_n)$, (or edges or vertices);

$$\omega_i = \frac{r_{i,k}}{n(n-1)...(n-i+1)} \tag{2.35}$$

$$r_{m,k} = (m-1)!\ (n-m+1)N_k(n,m-1) \ - \ m!N_k(n,\ m) \tag{2.36}$$

$$N_k(n,m) = \sum_{i=1}^{\left\lfloor \frac{n}{k} \right\rfloor} (-1)^i \binom{n-m+1}{i} \binom{n-k\cdot i}{n-m} \tag{2.37}$$

$t \in <0, +\infty)$
$\lfloor t \rfloor$ integer part of t.

In the variant without using the usability times, we obtain the definition in the following form:

A two-state complex CNI is called a vertex-"consecutive k out of n: F" (edge-"consecutive l out of m: F"), if it is faulty if and only if at least its k (l) successive ordered in the sequence of n (m) vertices (edges), i.e. V_1, V_2, ...,V_n (E_1, E_2,...,E_m) are faulty.

With respect to the structures, the reliability functions, assuming that failures are independent events and based on Eqs. 1.33–1.38, are given by the following formulas:

- for vertex-serial (edge-serial) CNIs:

$$R(t) = \prod_{i=1}^{n} VR_i(t), \tag{2.38}$$

$$(R(t) = \prod_{j=1}^{m} ER_j(t)), \tag{2.39}$$

and $t \in < 0, +\infty)$

- for vertex-parallel (edge-parallel) CNIs:

$$R(t) = 1 - \prod_{i=1}^{n} VF_i(t) \tag{2.40}$$

$$(R(t) = 1 - \prod_{j=1}^{m} EF_j(t), t \in < 0, +\infty); \tag{2.41}$$

and $t \in < 0, +\infty)$;

- for vertex-"k out of n" (edge-"l out of m") CNIs:

$$R(t) = 1 - \sum_{\substack{r_1, r_2, \ldots, r_n = 0 \\ r_1 + r_2 + \ldots + r_n \le l_v - 1}}^{1} [VR_i(t)]^{r_i} [VF_i(t)]^{1-r_i} \tag{2.42}$$

$$\left(R(t) = 1 - \sum_{\substack{r_1, r_2, \ldots, r_m = 0 \\ r_1 + r_2 + \ldots + r_m \le l_e - 1}}^{1} [ER_j(t)]^{r_j} [EF_j(t)]^{1-r_j} \right). \tag{2.43}$$

or

$$R(t) = \sum_{\substack{r_1, r_2, \ldots, r_n = 0 \\ r_1 + r_2 + \ldots + r_n \le \bar{l}_v}}^{1} [VF_i(t)]^{r_i} [VR_i(t)]^{1-r_i} \tag{2.44}$$

$$\left(R(t) = \sum_{\substack{r_1, r_2, \ldots, r_m = 0 \\ r_1 + r_2 + \ldots + r_m \le \bar{l}_e}}^{1} [EF_j(t)]^{r_j} [ER_j(t)]^{1-r_j} \right). \tag{2.45}$$

and $t \in < 0, +\infty)$, $\bar{l}_v = n - l_v$, $\bar{l}_e = n - l_e$

- for vertex-"consecutive k out of n: F" (edge-"consecutive l out of m: F") CNIs:

$$R_{k,l_v}(t) = \begin{cases} 1 - \prod\limits_{j=1}^{l_v} VF_j(t) & \text{for } k = l_v, \\ VR_{l_v}(t) R_{k,l_v - 1}(t) & \\ + \sum\limits_{i=1}^{k-1} VR_{l-i}(t) R_{k,l_v - i - 1}(t) \cdot \prod\limits_{j=l_v - i + 1}^{l_v} VF_j(t) & \text{for } k < l_v, \end{cases} \tag{2.46}$$

$$\left(R_{k,l_e}(t,u) = \begin{cases} 1 - \prod_{j=1}^{l_e} EF_j(t) & \text{for } k = l_e, \\ ER_{l_e}(t) R_{k,l_e-1}(t) \\ + \sum_{i=1}^{k-1} ER_{l_e-i}(t) R_{k,l_e-i-1}(t) \cdot \prod_{j=l_e-i+1}^{l_e} EF_j(t) & \text{for } k < l_e, \end{cases} \right) \tag{2.47}$$

where $l_v = 1,2,\ldots,n$ $l_e = 1,2,\ldots,m,$.

Similarly, one can find the characteristics of individual vertices or edges of the CNI.

In the particular case of the CNI model, defined following Figure 2.2, a definition of a two-state parallel-series system should be introduced, consistent with Guze (2014c, 2015); Kołowrocki (2004); Khan (2014); Kołowrocki and Soszyńska-Budny (2011). For this purpose, we adopt the following notations:

- n – number of system's components;
- E_{ij}, $i = 1,2,\ldots,k_m$, $j = 1,2,\ldots,l_i$– highlighted elements of CNIs;
- T_{ij} – independent random variables representing the lifespan of components E_{ij}, $i = 1,2,\ldots,k_m$, $j = 1,2,\ldots,l_i$(vertices or edges);
- $R_{ij}(t) = P(T_{ij} > t)$, $t \in< 0,+\infty)$ – reliability function of network elements of critical infrastructures E_{ij}, $i = 1,2,\ldots,k_m$, $j = 1,2,\ldots,l_i$(vertices or edges);
- $F_{ij}(t) = 1 - R_{ij}(t) = P(T_{ij} \leq t)$, $t \in< 0,+\infty)$– distribution of lifetimes T_{ij} of the network components E_{ij}, $i = 1,2,\ldots,k_n$, $j = 1,2,\ldots,l_i$(vertices or edges).

Moreover, we assume that the elements E_{ij}, $i = 1,2,\ldots,k_m$, $j = 1,2,\ldots,l_i$ (in the case of the considered network model defined in accordance with Figure 2.2) form parallel network subsystems and these in turn form the entire CNIs (in the vertex or edge sense).

A two-state system is called parallel-series if its lifetimes are given by the following relationship:

$$T = \min_{1\leq i\leq k_n} \{ \max_{1\leq j\leq l_i} T_{ij} \}. \tag{2.48}$$

Then the formula for the two-state reliability function of parallel-series systems is as follows (Kołowrocki, 2004; Khan, 2014; Kołowrocki and Soszyńska-Budny, 2011):

$$R_{k_n,l_1,\ldots,l_{k_n}j}(t) = \prod_{i=1}^{k_n} \left[1 - \prod_{j=1}^{l_i} F_{ij}(t) \right], t \in< 0,+\infty). \tag{2.49}$$

2.3.2 MULTI-STATE RELIABILITY MODELS OF CRITICAL NETWORK INFRASTRUCTURE

The proposed definitions assume that the reliability of the CNI deteriorates through the removal of edges or vertices. This allows for a simple description of the reliability states of the CNI using the number of removed elements (edges or vertices). The following proposition holds for CNIs.

The topological reliability state of a multi-state critical network infrastructure can be defined as the number of removed/unavailable vertices or edges, in which case:

best state z – all vertices/edges are available;
intermediate state u – selected vertices/edges are inoperable, but the transport network remains connected;
worst state 0 – after removing a certain number of vertices/edges, two connectivity components are created in the graph representing the network, i.e., the network is no longer connected.

Note: At time $t = 0$, we assume that the CNI is in the best state – z. This means that all nodes and arcs are accessible and the distances between nodes are as short as possible – both in terms of distance and time. Each "outage" of a node or arc then increases the distance (in units of distance and/or time) for at least one pair of network nodes, and the reliability of the transport network deteriorates, as defined in Definition 5.1. When two components of the transport network's connectivity are achieved, we observe a lack of connectivity between the network nodes located in the individual sets that constitute these components, and thus the transport network becomes unusable, as defined in Definition 5.1

Remark: In the case of a serial network, due to the existence of a bridge or a separating vertex, the above proposition indicates that they have only two states – the best and the worst.

For the network and its elements, we assume that $\{0,1,\ldots,z\}$ is an ordered set of their states, where state 0 is considered the worst, and state z is considered the best. Additionally, the fact that the CNI and its elements deteriorate over time, unless they are repaired, is taken into account. Under these assumptions and based on Eqs. (3.12 and 3.25), for multi-state CNI, the multi-state reliability functions of the elements (vertices or edges) and the entire network take the following form:

- reliabity function of:
  a vertex V_{i_V}, $i_V = 1,2,\ldots,n$ is described by the vector:

$$VR_{i_V}(t,\cdot) = [VR_{i_V}(t,0), VR_{i_V}(t,1),\ldots,VR_{i_V}(t,z)] \tag{2.50}$$

where

$$VR_{i_V}(t,u) = P(sv_{V_{i_V}}(t) \geq u | sv_{V_{i_V}}(0) = z) \tag{2.51}$$

is the probability that the vertex V_{i_v}, $i_v = 1,2,\ldots,n$ at time $t \in< 0,+\infty)$ is in one of the states from the subset of states $\{u,u+1,\ldots,z\}$, if at time $t = 0$ it was in the best state z;
an edge E_{i_e}, $i_e = 1,2,\ldots,m$ is described by the vector:

$$ER_{i_e}(t,\cdot) = [ER_{i_e}(t,0), ER_{i_e}(t,1),\ldots, ER_{i_e}(t,z)], \tag{2.52}$$

where

$$ER_{i_e}(t,u) = P(s_{e_{i_e}}(t) \geq u | s_{e_{i_e}}(0) = z) \tag{2.53}$$

is the probability of the event that the edge E_{i_e}, $i_e = 1,2,\ldots,m$, at time $t \in< 0,+\infty)$ is in one of the states from the subset of states $\{u,u+1,\ldots,z\}$, if at time $t = 0$ it was in the best state z

- the reliability function of a multi-state CNI is described by the vector:

$$\mathbf{R}(t,\cdot) = [\mathbf{R}(t,0), \mathbf{R}(t,1),\ldots, \mathbf{R}(t,z)] \tag{2.54}$$

where

$$\mathbf{R}(t,u) = P(s_{NT}(t) \geq u | s_{NT}(0) = z) \tag{2.55}$$

is the probability that the CNI at time $t \in< 0,+\infty)$ is in one of the states from the subset of states $\{u,u+1,\ldots,z\}$ provided that at time $t = 0$ it was in the best state z.

Analogously to Eqs. 1.26–1.29, the probability of the event that:

- vertex V_i, $i = 1,2,\ldots,n$ is in state u, provided that at time $t = 0$ it was in state z, is given by the vector:

$$p_{v_i}(t,\cdot) = [p_{v_i}(t,0), p_{v_i}(t,1),\ldots, p_{v_i}(t,z)] \tag{2.56}$$

where

$$\begin{gathered} p_{v_i}(t,u) = P(s_{v_i}(t) = u | s_{v_i}(0) = z) \\ t \in< 0,+\infty) u = 1,2,\ldots,z \end{gathered} \tag{2.57}$$

edge E_i, $i = 1,2,\ldots,n$ is in state u provided that at time $t = 0$ it was in state z is expressed by the vector:

$$p_{e_i}(t,\cdot) = [p_{e_i}(t,0), p_{e_i}(t,1),\ldots, p_{e_i}(t,z)] \tag{2.58}$$

where

$$\begin{gathered} p_{e_i}(t,u) = P(s_{e_i}(t) = u | s_{e_i}(0) = z) \\ t \in< 0,+\infty) u = 1,2,\ldots,z. \end{gathered} \tag{2.59}$$

However, the probability that the CNI is in state u, given that at time $t = 0$ it was in state z, is expressed by the vector

$$p(t,\cdot) = [p(t,0), p(t,1),\ldots, p(t,z)] \tag{2.60}$$

where

$$p(t,u) = P(s_{NT}(t) = u | s_{NT}(0) = z)$$
$$t \in < 0, +\infty)$$
$$u = 1,2,\ldots,z.$$

The introduction of the above concepts allows us to define the reliability structures of multi-state, ageing CNIs and their elements, as well as the parameters characterising them. Similarly to the two-state case, the following definitions obtained from Section 1.4.2 are true.

A multi-state critical network infrastructure is

- vertex-serial (edge-serial) if its time NT(u) in the subset of reliability states $\{u, u+1, \ldots, z\}$ is described by the relationship:

$$NT(u) = \min_{1 \le i \le n} \{VT_i(u)\} \tag{2.61}$$

$$(NT(u) = \min_{1 \le j \le m} \{ET_j(u)\}) \tag{2.62}$$

where $u = 1, 2 \ldots, z$.

- parallel in the vertex (edge) sense, if its time NT(u) in the subset of reliability states is described by the relationship:

$$NT(u) = \max_{1 \le i \le n} \{VT_i(u)\} \tag{2.63}$$

$$NT(u) = \max_{1 \le j \le m} \{ET_j(u)\}) \tag{2.64}$$

where $u = 1, 2, \ldots, z$

- "l_v out of n" ("l_e out of m") in the vertex (edge) sense, if its time $T(u)$ in the subset of reliability states $\{u, u+1, \ldots, z\}$ describes the relationship:

$$NT(u) = VT_{(n-l_v+1)}(u), l_v = 1, 2, \ldots, n \tag{2.65}$$

$$(NT(u) = ET_{(m-l_e+1)}(u),\ l_e = 1, 2, \ldots, m) \tag{2.66}$$

where $VT_{(n-l_v+1)}(u)$ $(ET_{(n-l_e+1)}(u))$ is the l_v-th (l_e-th) statistic of the highest degree in the sequence of lifetimes of network elements $VT_1(u)$,$VT_2(u)$,$ldots$,$VT_n(u)$, ($ET_1(u)$,$ET_2(u)$, …,$ET_m(u)$), $u = 1, 2, \ldots, z$.

- "consecutive l_v of n: F" in the vertex sense ("consecutive l_e of m: F" in the edge sense), if it does not reside in the subset of reliability states $\{u, u+1, \ldots, z\}$, if and only if at least l_v (l_e) of its n (m) ordered into a sequence of vertices (edges) V_1, V_2, …, V_n (E_1, E_2, …, E_m) does not reside in this subset.

We will say that a serial CNI resides in a subset of reliability states $\{u, u+1, \ldots, z\}$, if all of its elements reside in it. Furthermore, a parallel network resides in a subset of reliability states $\{u, u+1, \ldots, z\}$, if at least one of its elements resides in it. In contrast, a multi-state "l_v out of n" ("l_e out of m") network resides in a subset of reliability states $\{u, u+1, \ldots, z\}$ if at least l_v (l_e) of its n (m) elements reside in this subset.

Concerning the thus defined reliability structures of multi-state CNI, the coordinates of the reliability function vector are as follows:

- for a vertex-serial (edge-serial) CNI:

$$R(t,u) = \prod_{i=1}^{n} VR_i(t,u) \tag{2.67}$$

$$\left(R(t,u) = \prod_{j=1}^{m} ER_j(t,u)\right) \tag{2.68}$$

where $t \in <0, +\infty)$, $u = 1, 2, \ldots, z$

- for a vertex-parallel (edge-parallel) CNI:

$$R(t,u) = \prod_{i=1}^{n} VF_i(t,u) \tag{2.69}$$

$$\left(R(t,u) = 1 - \prod_{j=1}^{m} EF_i(t,u)\right) \tag{2.70}$$

where $t \in <0, +\infty)$, $u = 1, 2, \ldots, z$

- for a vertex - "l_v out of n" (edge - "l_e out of m") CNI:

$$R(t,u) = 1 - \sum_{\substack{r_1, r_2, \ldots, r_n = 0 \\ r_1 + r_2 + \cdots + r_n \leq l_v - 1}}^{1} [VR_i(t,u)]^{r_i} [VF_i(t,u)]^{1-r_i} \tag{2.71}$$

$$\left(R(t,u) = 1 - \sum_{\substack{r_1, r_2, \ldots, r_m = 0 \\ r_1 + r_2 + \cdots + r_m \leq l_e - 1}}^{1} [ER_i(t,u)]^{r_i} [EF_i(t,u)]^{1-r_i}\right) \tag{2.72}$$

or

$$R(t,u) = \sum_{\substack{r_1, r_2, \ldots, r_n = 0 \\ r_1 + r_2 + \ldots + r_n \leq \bar{l}_v}}^{1} [VF_i(t,u)]^{r_i} [VR_i(t,u)]^{1-r_i} \tag{2.73}$$

$$\left(R(t,u) = \sum_{\substack{r_1, r_2, \ldots, r_m = 0 \\ r_1 + r_2 + \ldots + r_m \leq \bar{l}_e}}^{1} [EF_i(t,u)]^{r_i} [ER_i(t,u)]^{1-r_i}\right) \tag{2.74}$$

where $t \in <0, +\infty)$, $u = 1, 2, \ldots, z$, $\bar{l}_v = n - l_v$, $\bar{l}_e = n - l_e$

- for vertex-"consecutive k out of n: F" (edge-"consecutive k out of m: F") CNI:

$$R_{k.l_v}(t,u) = \begin{cases} 1 - \prod_{j=1}^{l_v} VF_j(t,u) & \text{for } k = l_v, \\ VR_{l_v}(t,u)R_{k.l_v-1}(t,u) & \\ + \sum_{i=1}^{k-1} VR_{l-i}(t,u)R_{k.l_v-i-1}(t,u) \cdot \prod_{j=l_v-i+1}^{l_v} VF_j(t,u) & for\ k < l_v \end{cases} \tag{2.75}$$

$$\left(R_{k.l_e}(t,u) = \begin{cases} 1 - \prod_{j=1}^{l_e} EF_j(t,u) & \text{for } k = l_e, \\ ER_{l_e}(t,u)R_{k.l_e-1}(t,u) & \\ + \sum_{i=1}^{k-1} ER_{l_e-i}(t,u)R_{k.l_e-i-1}(t,u) \cdot \prod_{j=l_e-i+1}^{l_e} EF_j(t,u) & for\ k < l_e \end{cases} \right) \tag{2.76}$$

where $t \in < 0, +\infty)$, $u = 1, 2, \ldots, z$.

The basic parameters for the distinguished elements of the CNI, i.e., vertices V_i, $i = 1, 2, \ldots, n$, and edges $E_j, j = 1, 2, \ldots, m$, are defined as follows:

the expected value of the residence time in the subset of states $\{u, u+1, \ldots, z\}$ is given by the following formula:

$$M_i(u) = \int_0^\infty VR_i(t,u)dt \tag{2.77}$$

$$M_j(u) = \int_0^\infty ER_j(t,u)dt \tag{2.78}$$

standard deviation of the residence time in the subset of states $\{u, u+1, \ldots, z\}$:

$$\sigma_i(u) = \sqrt{N_i(u) - [M_i(u)]^2} \tag{2.79}$$

$$\sigma_j(u) = \sqrt{N_j(u) - [M_j(u)]^2} \tag{2.80}$$

where

$$N_i(u) = 2\int_0^\infty tVR_i(t,u)dt \tag{2.81}$$

$$N_j(u) = 2\int_0^\infty tER_j(t,u)dt \tag{2.82}$$

Based on Eqs. 1.39–4.5, the selected parameters of the CNI are given as follows:

the expected value of the residence time in the subset of states $\{u, u+1, \ldots, z\}$ is given by the following formula:

$$M(u) = \int_0^\infty \mathbf{R}(t,u)dt \tag{2.83}$$

standard deviation of the time spent in a subset of states:

$$\sigma(u) = \sqrt{N(u) - [M(u)]^2} \tag{2.84}$$

where

$$N(u) = 2\int_0^\infty t\mathbf{R}(t,u)dt. \tag{2.85}$$

2.3.3 MIXED RELIABILITY MODELS OF CRITICAL NETWORK INFRASTRUCTURE

When considering the reliability of multi-state ageing CNI, we should additionally consider the situation in which the CNI is a multi-state technical system and its components, i.e., vertices and edges, are two-state systems. In this case, for the CNI, as in Section 2.3.2, we assume that it is an ordered set of its states, where state 0 is considered the worst and z is the best. However, for the nodes and arcs of this network, it is a two-element set – they operate perfectly (state 1) or not at all (state 0). Assuming that the CNI is not subject to repairs and its condition deteriorates over time, the multi-state reliability functions are expressed as follows:

- the reliability function of two-state nodes V_i, $i = 1,2,\ldots,n$ and network arcs E_i, $j = 1,2,\ldots,m$ are described by the following formulas:

$$VR_i(t,\cdot) = [VR_i(t,0), VR_i(t,1)] \tag{2.86}$$

where

$$VR_{i_v}(t,u) = P(s_{V_{i_v}}(t) \geq u | s_{V_{i_v}}(0) = z) \tag{2.87}$$

is the probability of the event that a vertex V_i, at time t, $t \in <0,\infty)$ is in one of the states from the subset of states $\{0,1\}$, provided that at time $t = 0$ it was in the best state 1

and

$$ER_j(t,\cdot) = [ER_j(t,0), ER_j(t,1)] \tag{2.88}$$

where

$$ER_{i_e}(t,u) = P(s_{e_{i_e}}(t) \geq 0 | s_{e_{i_e}}(0) = 1) \tag{2.89}$$

is the probability of the event that the edge E_{i_e} at time $t \in <0,+\infty)$ is in one of the states from the subset of states $\{0,1\}$, provided that at time $t = 0$ it was in the best state 1;

the following vector describes the reliability

$$\mathbf{NR}(t,\cdot) = [\mathbf{R}(t,0), \mathbf{R}(t,1)$$

where $\mathbf{R}(t,u) = P(s_{NT}(t) \geq u | s_{NT}(0) = z)$ i at time $t \in <0,\infty)$ is in one of the states fro $1,\ldots,z\}$,if at time $t = 0$ it was in the best stat

The probability that edge E_i, $i = 1,2,\ldots,m$, is $t = 0$ it was in state 1, is given by the vector:

$$p_{e_i}(t,\cdot) = [p_{e_i}(t,0), p_{e_i}(t,1)] \tag{2.91}$$

where

$$p_{e_i}(t,u) = P(s_{e_i}(t) = u | s_{e_i}(0) = 1) \tag{2.92}$$

and $t \in <0,+\infty)$, $u = \{0,1\}$.

Similarly, the probability of the event that vertex V_i, $i = 1,2,\ldots,n$, is in state u, provided that at time $t = 0$ it was in state 1, is given by the vector:

$$[p_{v_i}(t,\cdot) = [p_{v_i}(t,0), p_{v_i}(t,1)] \tag{2.93}$$

where

$$p_{v_i}(t,u) = P(s_{v_i}(t) = u | s_{v_i}(0) = 1) \tag{2.94}$$

and $t \in <0,+\infty)$, $u = \{0,1\}$.

However, the probability that the CNI is in state u, given that at time $t = 0$ it was in the best state z, is given by the vector:

$$p(t,\cdot) = [p(t,0), p(t,1),\ldots, p(t,z)] \tag{2.95}$$

where

$$p(t,u) = P(s(t) = u | s(0) = z) \tag{2.96}$$

and $t \in <0,+\infty)$,$u = 1,2,\ldots,z$.

The introduction of the above concepts allows us to define the reliability structures of multi-state, ageing CNIs and their elements, as well as the parameters characterising these networks. Analogously to the cases described in Sections 2.3.1 and 2.3.2, the following definitions apply to a multi-state transportation network composed of two-state arcs and nodes. A multi-state CNI with binary components (nodes and arcs) is

- vertex-serial (edge-serial) if its time $NT(u)$ in the subset of reliability states $\{u,u+1,\ldots,z\}$ is described by the relationship:

$$NT(u) = \min_{1 \leq i \leq n} \{VT_i\} \tag{2.97}$$

$$(NT(u) = \min_{1 \leq j \leq m} \{ET_j\})u = 1,2\ldots,z. \tag{2.98}$$

(edge-parallel), if its time $NT(u)$ in the subset of reliability ...cribed by the relationship:

$$NT(u) = \max_{1 \leq i \leq n} \{VT_i\} \tag{2.99}$$

$$NT(u) = \max_{1 \leq j \leq m} \{ET_j\}) \tag{2.100}$$

where $u = 1,2,\ldots,z$

- vertex-"l_v out of n" (edge-"l_e out of m"), if its time $NT(u)$ in the subset of reliability states $\{u,u+1,\ldots,z\}$ describes the relationship:

$$NT(u) = VT_{(n-l_v+1)}, l_v = 1,2,\ldots,n \tag{2.101}$$

$$(NT(u) = ET_{(m-l_e+1)},\ l_e = 1,2,\ldots,m) \tag{2.102}$$

where $VT_{(n-l_v+1)}$ $(ET_{(n-l_e+1)})$ is the l_v-th (l_e-th) statistic of the highest degree in the sequence of lifetimes of network elements VT_1,VT_2,$ldots$,VT_n, (ET_1,ET_2, ...,ET_m), $u = 1,2,\ldots,z$.

- vertex-"consecutive l_v of n: F" (edge-"consecutive l_e of m: F"), if it does not reside in the subset of reliability states $\{u,u+1,\ldots,z\}$, if and only if at least l_v (l_e) of its n (m) ordered into a sequence of vertices (edges) V_1, V_2, ..., V_n (E_1, E_2, ..., E_m) does not reside in this subset.

We say that a serial CNI resides in a subset of reliability states $\{u,u+1,\ldots,z\}$, if all of its elements reside in it. Furthermore, a parallel network resides in a subset of reliability states $\{u,u+1,\ldots,z\}$, if at least one of its elements resides in it. In contrast, a multi-state "l_v out of n" ("l_e out of m") network resides in a subset of reliability states $\{u,u+1,\ldots,z\}$ if at least l_v (l_e) of its n (m) elements reside in this subset.

Concerning the thus defined reliability structures of multi-state CNI with two-state components, the coordinates of the reliability function vector are as follows:

- for a vertex-serial (edge-serial) CNI:

$$R(t,u) = \prod_{i=1}^{n} VR_i(t) \tag{2.103}$$

$$(R(t,u) = \prod_{j=1}^{m} ER_j(t)) \tag{2.104}$$

where $t \in <0,+\infty)$, $u = 1,2,\ldots,z$

- for a vertex-parallel (edge-parallel) CNI:

$$R(t,u) = \prod_{i=1}^{n} VF_i(t) \tag{2.105}$$

$$(R(t,u) = 1 - \prod_{j=1}^{m} EF_i(t)) \tag{2.106}$$

where $t \in <0,+\infty)$, $u = 1,2,\ldots,z$

- for a vertex-"l_v out of n" (edge-"l_e out of m") CNI:

$$R(t,u) = 1 - \sum_{\substack{r_1,r_2,\ldots,r_n=0 \\ r_1+r_2+\cdots+r_n \le l_v-1}}^{1} [VR_i(t)]^{r_i}[VF_i(t)]^{1-r_i} \tag{2.107}$$

$$\left(R(t,u) = 1 - \sum_{\substack{r_1,r_2,\ldots,r_m=0 \\ r_1+r_2+\cdots+r_m \le l_e-1}}^{1} [ER_i(t)]^{r_i}[EF_i(t)]^{1-r_i} \right) \tag{2.108}$$

or

$$R(t,u) = \sum_{\substack{r_1,r_2,\ldots,r_n=0 \\ r_1+r_2+\ldots+r_n \le \bar{l}_v}}^{1} [VF_i(t,u)]^{r_i}[VR_i(t,u)]^{1-r_i} \tag{2.109}$$

$$\left(R(t,u) = \sum_{\substack{r_1,r_2,\ldots,r_m=0 \\ r_1+r_2+\ldots+r_m \le \bar{l}_e}}^{1} [EF_i(t,u)]^{r_i}[ER_i(t,u)]^{1-r_i} \right) \tag{2.110}$$

where $t \in <0,+\infty)$, $u = 1,2,\ldots,z$, $\bar{l}_v = n - l_v$, $\bar{l}_e = n - l_e$

- for vertex-"consecutive k out of n: F" (edge-"consecutive k out of m: F") CNI:

$$R_{k,l_v}(t,u) = \begin{cases} 1 - \prod_{j=1}^{l_v} VF_j(t,u) & \text{for } k = l_v, \\ VR_{l_v}(t,u)R_{k,l_v-1}(t,u) & \\ + \sum_{i=1}^{k-1} VR_{l-i}(t,u)R_{k,l_v-i-1}(t,u) \cdot \prod_{j=l_v-i+1}^{l_v} VF_j(t,u) & for\, k < l_v \end{cases} \tag{2.111}$$

$$\left(R_{k.l_e}(t,u) = \begin{cases} 1 - \prod_{j=1}^{l_e} EF_j(t,u) & \text{for } k = l_e, \\ ER_{l_e}(t,u)R_{k.l_e-1}(t,u) & \\ + \sum_{i=1}^{k-1} ER_{l_e-i}(t,u)R_{k.l_e-i-1}(t,u) \cdot \prod_{j=l_e-i+1}^{l_e} EF_j(t,u) & for\, k < l_e \end{cases} \right) \tag{2.112}$$

where $t \in <0,+\infty)$, $u = 1,2,\ldots,z$.

The basic parameters for the distinguished elements of the CNI, i.e., vertices V_i, $i = 1,2,\ldots,n$, and edges $E_j, j = 1,2,\ldots,m$, are defined as follows:

the expected value of the residence time in the subset of states $\{u, u+1, \ldots, z\}$ is given by the formula:

$$M_i(u) = \int_0^\infty VR_i(t)dt \tag{2.113}$$

$$M_j(u) = \int_0^\infty ER_j(t)dt \tag{2.114}$$

standard deviation of the residence time in the subset of states $\{u, u+1, \ldots, z\}$:

$$\sigma_i(u) = \sqrt{N_i(u) - [M_i(u)]^2} \tag{2.115}$$

$$\sigma_j(u) = \sqrt{N_j(u) - [M_j(u)]^2} \tag{2.116}$$

where

$$N_i(u) = 2\int_0^\infty tVR_i(t)dt \tag{2.117}$$

$$N_j(u) = 2\int_0^\infty tER_j(t)dt \tag{2.118}$$

2.3.4 METHODS AND ALGORITHMS FOR ANALYSIS TOPOLOGICAL RELIABILITY

Based on the definitions presented in Section 2.3, two algorithms can be proposed for assessing the reliability of CNI by identifying their reliability structures. Below are two flowcharts describing the general steps of the individual algorithms (see Figures 2.4 and 2.5). Identifying the structures allows us to find the basic reliability parameters for CNIs.

Consider a CNI represented by an undirected graph $G(V,E)$. Furthermore, consider the reliability structures defined by the author in previous sections. The steps of the individual algorithms for identifying reliability structures, both edge and vertex, are presented in Algorithms 2.5 and 2.6. They correspondence to flowcharts in Figures 2.4 and 2.5. Referring to the flowchart shown in Figure 2.4, we have Algorithm 2.5

In the case of structures relating to CNI nodes, as shown in Figure 3.2, we have Algorithm 2.6.

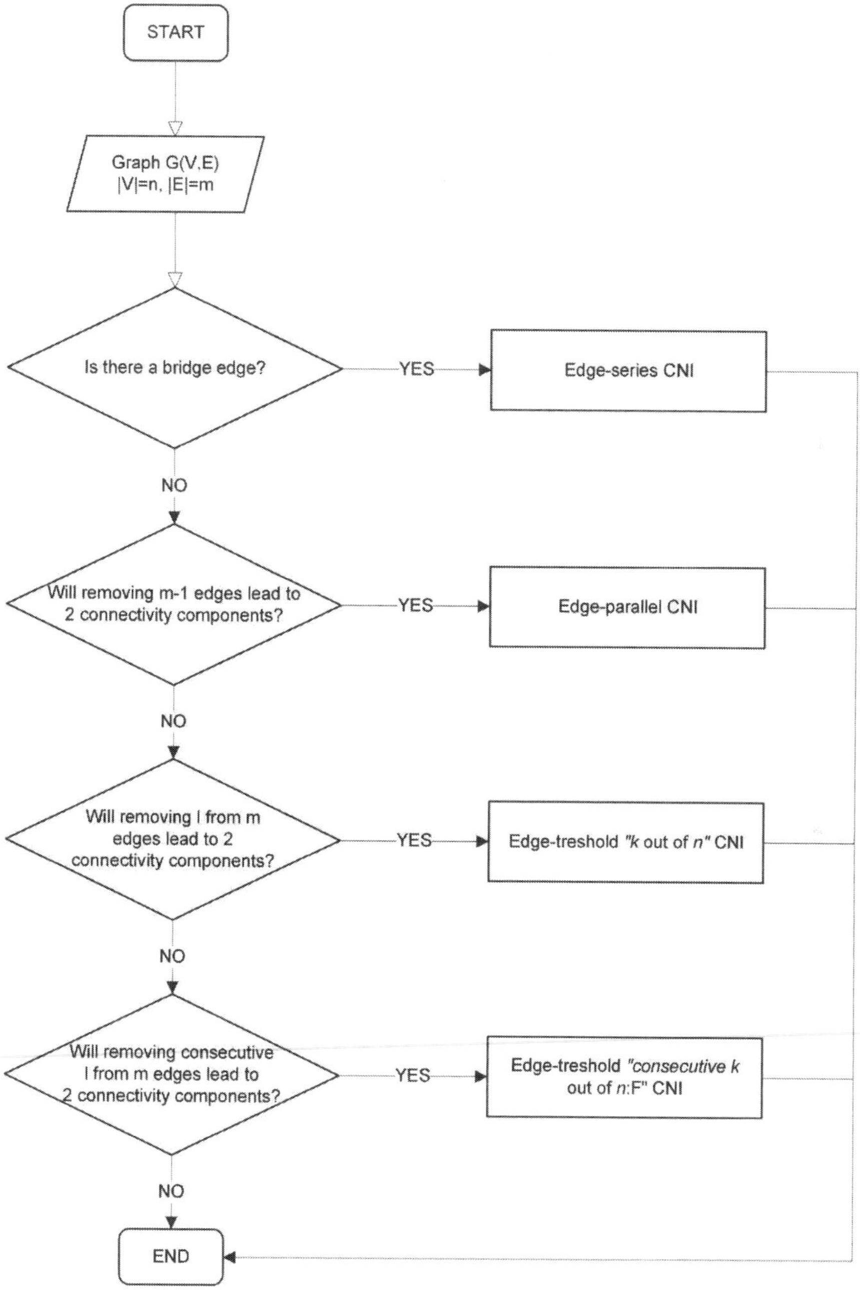

Figure 2.4 Block diagram of algorithm for CNI edges structures

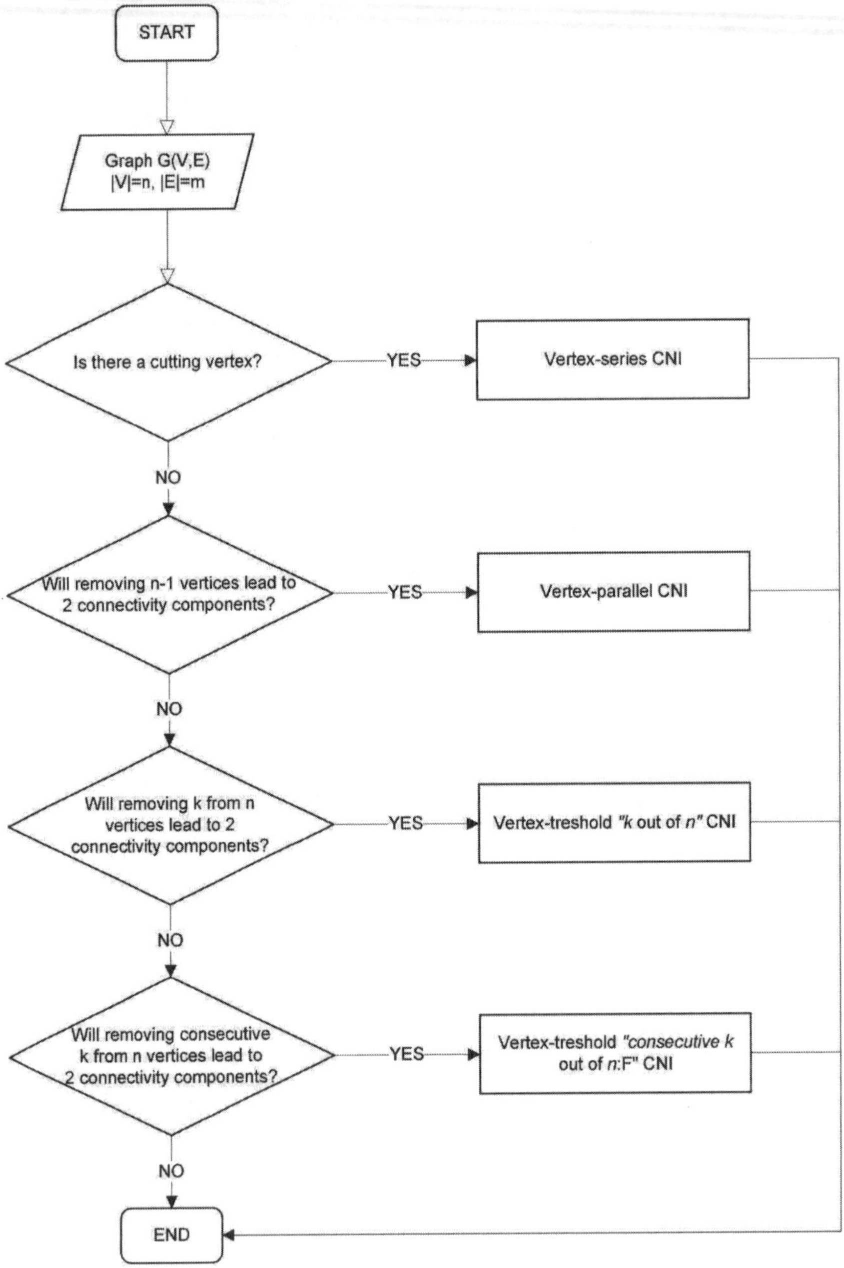

Figure 2.5 Block diagram of algorithm for CNI nodes structures

Algorithm 2.5: Identifying Edge Reliability Structures

1. Checking whether a bridge exists in the graph using Algorithm A.2;
2. Checking whether there are $m - 1$ edges whose removal would result in at least two connectivity components using Algorithm A.3;
3. Checking whether there are k of the m damaged edges, which is accomplished by searching for the edge-dominant set of the lowest cardinality, performed by a modified version of Algorithm A.4;
4. Checking whether there are another k of the m available damaged edges, which is accomplished by a modified version of Algorithm A.4.

Algorithm 2.6: Identifying Node Reliability Structures

1. Checking whether a cutting vertex exists in the graph, which is performed using Algorithm A.5;
2. Checking whether there are n-1 vertices whose removal will result in the creation of at least two connectivity components, performed by Algorithm A.6;
3. Checking whether any k out of n available vertices exist, which is performed by Algorithm A.6;
4. Checking whether another k out of n available vertices exist, which is performed by a modified version of Algorithm A.6.

Detailed algorithms implementing the individual steps of the flowchart are presented in Algorithms A.2–A.4.

The proposed algorithms are not the only ones implementing the individual stages of the procedures for identifying the reliability structures of CNIs. In the future, especially concerning large, complex networks, the possibility of using algorithms with the lowest possible computational and memory complexity should be considered. Nevertheless, it should be noted that the proposed methods for modelling the reliability of CNIs are based mainly on solutions to graph theory problems, which are classified as NP-complete or NP-hard (non-polynomial deterministic; Cormen et al., 2009; Fink et al., 1990; Haynes et al., 1998, 2000; Leeuwen, 1986; Zhan and Noon, 1998).

The block diagrams presented in Figures 3.1 and 3.2 also facilitate a comprehensive characterisation of the reliability structures of the entire network, taking into account both the vertex and edge approaches. This approach leads to a description in which we obtain a joint vertex-edge structure. Then, the reliability structures of the CNI should be extended according to the definition below, taking into account the concepts introduced in earlier definitions.

CRITICAL NETWORK INFRASTRUCTURES VERTEX–EDGE RELIABILITY STRUCTURE

A critical network infrastructure is said to have a vertex-edge reliability structure of the type consistent with the structures of the network nodes described as vertex-serial, vertex-parallel, vertex-and the arcs described in Definitions 5.3, 5.5, 5.7, and 5.9.

In other words, when considering joint models of reliability structures to describe a CNI reliability, the following terms are used:

1. series – series;
2. series – parallel;
3. series – "*k* out of *m*";
4. series – "consecutive *k* out of *m*: F";
5. parallel – series;
6. parallel – parallel;
7. parallel – "*k* out of *m*";
8. parallel – "consecutive *k* out of *m*: F";
9. "*k* out of *n*"– series;
10. "*k* out of *n*" – parallel;
11. "*k* out of *n*" – "*k* out of *m*";
12. "*k* out of *n*" – "consecutive *k* out of *m*: F";
13. "consecutive *k* out of *n*: F" – serial;
14. "consecutive *k* out of *n*: F" – parallel;
15. "consecutive *k* out of *n*: F" – "*k* out of *m*";
16. "consecutive *k* out of *n*: F" – "consecutive *k* out of *m*: F".

We assume that in the above definition and the above index, the type before the hyphen denotes the structure of the transport network in the vertex sense, and the type after the hyphen denotes the structure of the transport network in the edge sense. At the same time, to shorten the notation in the case of structures a), f), k) and p), we will call the structure of the transport network serial, parallel, "*k* out of *n*" or "*k* out of *m*", "consecutive *k* out of *n*: F" or "consecutive k out of m: F."

2.4 CASE STUDIES

This section describes three case studies. First is about US pipeline network analysis. Second case analyse Saudi Arabia pipeline network. Third case is about city transportation network.
We take into account the two cases about pipelines in the United States and Saudi Arabia because of their importance in everyday life and society functioning. Generally, the total length of the global transmission pipeline network is around 2.28 million km. The crude oil pipelines constitute 18%, petroleum products – 12%, natural gas

pipelines – 64%, and natural gas liquid (NGL) – 6% (according to GlobalData Energy).

2.4.1 USA PIPELINES MODEL

Firstly, we take into account US main pipelines. The total length is around 815,358 km, where the major pipelines are as follows:

Cochin Pipeline – 1,900 miles
Keystone Pipeline – 1,853 miles
Appalachia to Texas Express Pipeline – 1,205 miles
Dakota Access Pipeline – 1,172 miles
Overland Pass Pipeline – 1,035 miles
Southern Hills Pipeline – 940 miles
Platte Pipeline – 932 miles
Gray Oak Pipeline – 850 miles
Alberta Clipper Pipeline – 786 miles
Express Pipeline – 785 miles
Pony Express Pipeline – 760 miles
Energy Transfer Crude Oil Pipeline – 754 miles
EPIC Crude Pipeline – 730 miles
Spearhead Pipeline – 650 miles

Figure 2.6 presents graph model of US main pipelines.

Now, we can find the topological reliability structure under theory introduced in Section 2.3.1. Considering the graph in Figure 2.6, we can see that it has many separating vertices. This implies that it is a vertex-serial network in the sense of topological reliability.

We proceed similarly to above. Since the network contains many bridges, we can conclude that it is an edge-serial network.

Taking into account the results of reliability modelling in the vertex and edge senses, we generally obtain a serial-serial structure for this network, indicating that it is serial.

When we know what type of structure this network has, we can give the reliability functions tuple: nodes, edges and whole network, according to formulae 2.38 and 2.39 in two-state model or 2.50 and 2.52 in multi-state approach.

2.4.2 SAUDI ARABIA PIPELINES MODEL

Let's take into account the pipelines in Saudi Arabia. A total length of pipeline network is over 20,000 km (active and closed), where the major pipelines are:

East-West Crude Oil Pipeline – with 1,201 km
Abqaiq-Yanbu Oil Line I and II – 1,204 km
Bahrain-Saudi Arabia Oil Pipeline – 115 km.

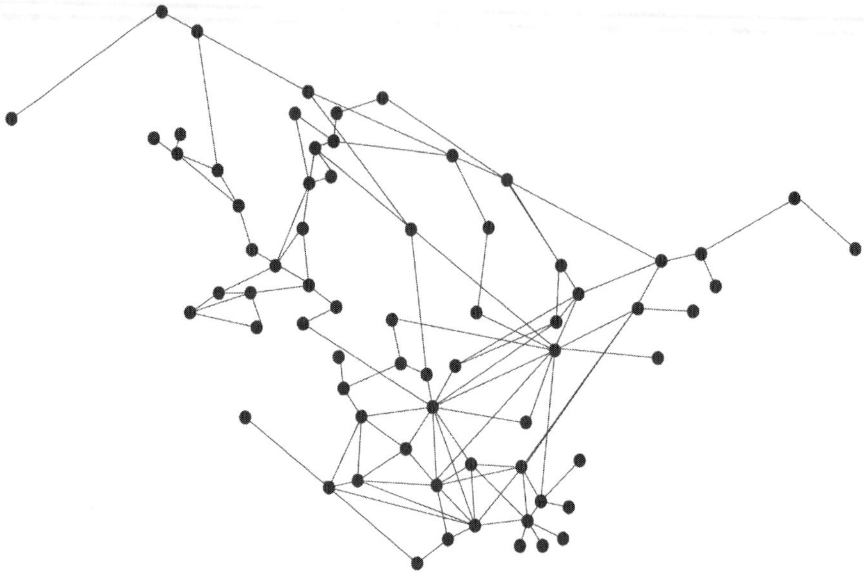

Figure 2.6 Graph model of US pipeline network

According to data of US Energy Information Administration Saudi Arabia crude oil production is at a current level of 9.12*M* barrels, down from 9.32*M* last month and down from 9.47*M* in 2024. This is a change of −2.15% from last month and −3.70% from one year ago. Table 2.3 shows volume of oil and others products transported through the Strait of Hormuz.

It is crucial to modelling and analyse reliability and vulnerability of this pipeline network. Because if the Strait of Hormuz will be closed, the Saudi Arabia pipelines give possibility to use Abqaiq-Yanbu Oil Line I and II for sending the oil to Red Sea port Yandu.

In this chapter, we only show the graph theory model and topological reliability model (see Figure 2.7 of the main pipelines in Saudi Arabia. In Section 4.4 the vulnerability analysis is done).

Considering the graph model of the Saudi Arabia pipeline network and according to definitions vertex and edge sense topological reliability, we identify that this network is vertex-serial and edge-serial . There are two separate vertices and many edges, which are bridges in the graph representing this network (see Figure 2.6. It means that pipeline network is serial-serial, i.e., serial).

2.4.3 MODELLING THE RELIABILITY OF ROADS WITH DIGRAPHS

This subsection describes graph theory model of the crossroads (it can be found in Gdynia, Poland). This approach uses the theory introduced in Subsection 2.3.1 to describe these transportation network components' reliability structures (terms intro-

Table 2.3
Volume of Crude Oil, Condensate, and Petroleum Products Transported Through the Strait of Hormuz, 2020 – 1Q2025 (Million Barrels Per Day)

	2020	2021	2022	2023	2024	1Q2025
Total oil flows – Strait of Hormuz	19.1	19.4	21.4	21.4	20.3	20.1
Crude oil and condensate	14.3	14.4	16.0	15.5	14.3	14.2
Petroleum products	4.8	5.0	5.5	5.8	5.9	5.9
World maritime oil trade	71.4	72.6	74.3	76.0	75.5	75.7
World total petroleum and other liquids consumption	91.0	96.6	99.5	101.8	102.7	102.1

Source: Own work based on U.S. Energy Information Administration, Short-Term Energy Outlook, June 2025, and U.S. Energy Information Administration analysis based on Vortexa tanker tracking

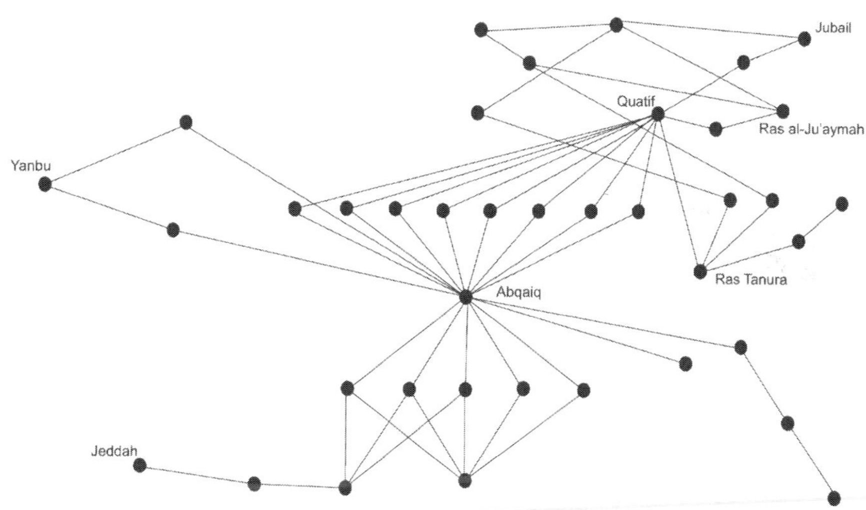

Figure 2.7 Graph model of Saudi Arabia pipeline network

duced in 2.2. Figure 2.8a presents a crossroad's physical model as a transportation network representation.

Figure 2.8b presents the adequate graph theory model. We use the directed graph, i.e., a graph where edges have a specific direction indicated by arrows pointing from one node to another (pair of ordered vertices). One directed edge of the graph corresponds to one lane connecting the vertices of the transportation network. Thus, there are following edges between the particular vertices:

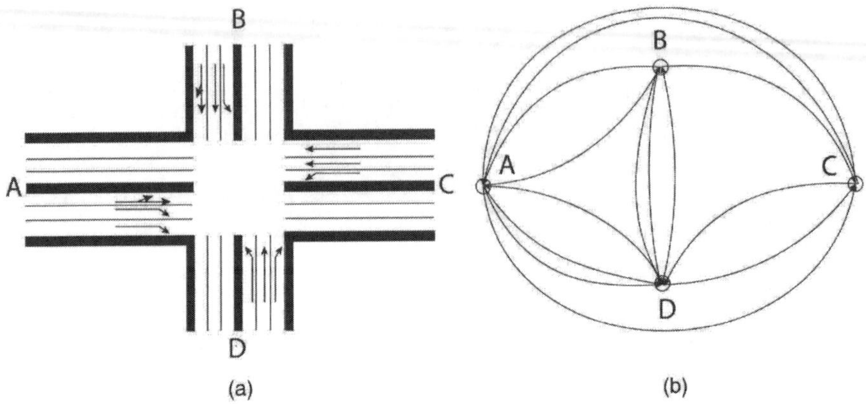

Figure 2.8 Directed graph model of crossroad

from A to C: one edge;
from C to A: two edges;
from A to B: one edge;
from B to A: one edge;
from A to D: two edges;
from D to A: one edge;
from B to C: one edge;
from B to D: two edges;
from D to B: one edge;
from C to D: one edge;
from D to C: one edge.

This transportation network model allows the use of a structure described in Subsection 2.3.1. The selection of the structure type depends on the assumption about when a connection between a transportation network's nodes is inoperable.
For example, the case of the nodes A and D can be a serial structure or a parallel structure. The first type is used when we assume that the connection between these nodes is reliable when all lanes are operable. The second structure describes the case in which even one traffic lane gives the connection between selected nodes.

3 Analysis of Critical Network Infrastructure

This chapter first presents a general approach to analysing critical infrastructures' reliability and vulnerability. It also includes basic knowledge about threats, risk assessment, mitigation, and ensuring business continuity. Then, the content focuses on describing new or modified tools for analysing the issues mentioned earlier. The chapter ends with case studies.

3.1 GENERAL APPROACH AND BASIC NOTATIONS FOR CRITICAL NETWORK INFRASTRUCTURES ANALYSIS

The reliability models of critical network infrastructures (CNIs) introduced in Chapter 2 provide us with tools for conducting analyses on three levels: structural, functional, and decision-making. Parameters describing network reliability characteristics enable us to determine, for example, the expected value of lifetime or its standard deviation. The previously introduced fundamental concepts of graph theory are used not only for reliability models but also to define and introduce new parameters related to the idea of domination in graphs. These, in turn, are used to analyse the vulnerability or resilience of CNI. Moreover, the fundamentals of game theory provide the basis for decision analysis tools.

Typical methods for analysing technical systems in terms of reliability or resilience to various events include event trees and fault trees. However, they are not the primary analytical methods presented in this book. As previously mentioned, after identifying reliability structures, the parameters defining these structures can be used, as introduced in Chapter 2. It should be assumed that the mean lifetime or the mean time being in reliable states constitutes the basic reliability characteristics of CNIs. The same applies to standard deviation. In probability theory, the mean is a measure of centrality, and the standard deviation is a measure of dispersion. These characteristics enable a fundamental analysis of the performance of any technical system, particularly CNIs.

This chapter presents new methods and approaches for analysing CNI. They are based on the graph and game theories.

3.2 METHODS AND ALGORITHMS FOR ANALYSIS

First of all, we introduce some parameters to help providing an analysis of CNI. These parameters base on the domination in graphs. There are many types of dominating sets and, therefore, their corresponding domination numbers (Harrary, 1969;

DOI: 10.1201/9781003661450-3

Haynes et al., 1998). They are often determined concerning a selected family of graphs. Hence, in this work, we use three types of dominating sets: *regular*, connected, independent. Based on Harrary (1969) and Haynes et al. (1998) we present the definitions of these sets and numbers.

The set $D \subseteq V(G)$ is called a *dominating set* of a graph G if every vertex $v \in V$ belongs to this set or has a neighbour in it (i.e., a set in which every vertex $x \in V(G) - D$ has a neighbour in D). At the same time, the size of the smallest dominating set of a graph G is called the *domination number* of this graph and is denoted by $\gamma(G)$.

According to Walikar et al. (1979) and Guze (2019a), the notions of a connected dominating set and a connected dominating number are explained below.

A set $D_C \subseteq V(G)$ is called a *connected dominating set* if every vertex of $V \backslash D_c$ is adjacent to a vertex in D_C and the subgraph induced by D_C is connected. The size of the smallest connected dominating set in a graph G is called the connected domination number and is denoted by $\gamma_C(G)$.

A set $D_I \subseteq V(G)$ is called an *independent dominating set* of a graph G if the set D_I is independent and every vertex not in D_I is adjacent to a vertex in this set. The size of the smallest independent dominating set is called the independent dominating number and is denoted by $\gamma_I(G)$.

The concept of an independent set describes a set of independent vertices, i.e., vertices that are not adjacent to each other.

For the vertex-weighted graphs defined in Section 2.1.1, it is possible to define the numbers:

- weighted regular domination,
- weighted connected domination,
- weighted independent domination.

Detailed definitions of these numbers are as follows: The *weighted domination number* $\gamma_w(G)$ of the weighted graph (G, w_v) is the smallest weight $w(D) = \sum_{v \in D} w(v)$ of all vertices belonging to the dominating set $D \subseteq V(G)$. Furthermore, the *weighted connected domination number* $\gamma_{Cw}(G)$ of a weighted graph (G, w_v) is the smallest weight $w(D_C) = \sum_{v \in D_C} w(v)$ of all vertices belonging to the set $D_C \subseteq V(G)$. The *weighted independent domination number* $\gamma_{Iw}(G)$ of a weighted graph (G, w_v) is defined as the smallest weight $w(D_I) = \sum_{v \in D_I} w(v)$ of all vertices belonging to the set $D_I \subseteq V(G)$.

Therefore, it is justified to use domination in graphs for research on CNI. When we talk about the networks, the dominating sets contain those network nodes that absorb the highest traffic loads or should be under the special supervision of traffic/flow/energy/data control system operators. The rationale behind this is that the greedy algorithm (Algorithms 3.1 and 3.2) for searching the dominating set selects the nodes (vertices) with the highest degree (the largest set of neighbours). The steps of these algorithms are presented below based on, among others, the research results contained in Haynes et al. (1998) and Leeuwen (1986).

The above algorithms do not always return the minimum dominating set. This is because all problems involving dominating sets are in NP (nondeterministic

Algorithm 3.1: Minimal Dominating Set 1

1. Let $V = \{1,\ldots,n\}$, and $D = \emptyset$.
2. At each step, greedily add a new vertex to D until it forms a dominating set.
3. Node j is called covered if $j \in D$ or if any neighbour of j belongs to D. A vertex that is not covered is called uncovered.
4. At each step, insert into D the lowest-indexed vertex that covers the largest number of uncovered vertices.
5. Stop when all vertices are covered.

Algorithm 3.2: Minimal Dominating Set 2

1. Set $S := 0$; all vertices are white;
2. As long as white nodes exist, execute
   a. Choose $v \in \{x | w(x) = \max_{u \in V} \{w(u)\}\}$;
   b. $S := S \cup \{v\}$; color the vertices in S black and their neighbours gray;
3. End until.

polynomial time - is the set of decision problems solvable in polynomial time by a nondeterministic Turing machine). Therefore, we propose a modification of Algorithms 3.1 and 3.2 below to create a single algorithm for finding the minimum dominating set. The input for the new algorithm is a connected, simple, non-empty graph. Algorithm 3.3, based on the two previous Algorithms 3.1 and 3.2, can then be used to find the minimum dominating set.

Note: The set S is a dominating set, but it is not always the minimal set. Therefore, we check whether there are vertices in this set that dominate the same vertices and, at the same time, their neighbours are dominated by other vertices from S. If so, then such a vertex can be removed. By repeating this operation for all vertices in S and removing those that do not dominate the non-dominated vertices, we obtain the minimal dominating set.

One application of domination in graphs is to solve a problem posed by Fink et al. (1990). He decided to ask about the sensitivity of a computer/communication network to connection failures. He considered the possible situation in which someone does not know which nodes in the network act as transmitters, but knows that a set of such nodes can belong to the minimal dominating set in the graph representing the structure of the network. This posed the question of the minimum number of communication links that must be severed so that at least one additional transmitter is necessary to ensure communication with all nodes. This led to the proposal of a new parameter called the graph bondage number defined as follows.

The *bondage number* $b(G)$ of a non-empty graph G is defined as the smallest number of edges in a graph whose removal would increase the domination number in G, i.e., $\gamma(G-E) > \gamma(G)$.

Algorithm 3.3: Minimum Dominating Set – Guze

1. Set $S := 0$; all vertices are white;
2. As long as there are white nodes, execute
   a. Choose $\upsilon \in \{x|w(x) = \max_{u \in V} \{w(u)\}\}$ and has the largest number of white neighbours;
   b. $S := S \cup \{v\}$; we color vertices in S black and their neighbours gray;
3. end until;
4. return S;
5. if {in S there are vertices that dominate the same vertices};
6. then {remove the vertex from S that only dominates vertices dominated by other vertices in S; return $D = S\{v\}$ };
7. else $D = S$;
8. return D.

In other words, the bondage number describes the smallest set of edges whose removal from a graph G increases the domination number in that graph. However, this definition of the bondage number in a graph is not entirely appropriate due to the lack of a condition for maintaining the network's connectivity. Therefore, the definition of the new bondage number in a graph G is given below:

The *bondage connected number* $b_c(G)$ of a non-empty connected graph G is defined as the smallest number of edges in graph E whose removal will increase the domination number in graph G, i.e., $\gamma(G-E) > \gamma(G)$ and graph $G-E$ is connected.

In turn, when we consider the case of a connected domination number, we can give the following definition:

A *connected bondage number* $cb(G)$ of a non-empty, connected graph G is the least set of edges E for which $\gamma_C(G-E) > \gamma_C(G)$ holds.

We also introduce the weighted connected number, which help you provide more precisely analysis of CNIs represented by weighted graphs (vertex-weighted).

The *weighted bondage connected number* $wb_c(G)$ of a non-empty graph G is defined as the smallest cardinality among all edge sets for which $\gamma_w(G-E) > \gamma_w(G)$ and at the same time the graph $G-E$ is connected.

Another concept related to domination in graphs is called *edge subdivision*. The operation itself is defined as follows (Haynes et al., 2000; Bhattacharya and Vijayakumar, 2002).

We say that an edge $uv \in E$ is divisible if it is possible to remove this edge and add a new vertex (called the split vertex) in its place, along with two new edges: ux and xv.

The above definition is based on the assumption that only a single division of an edge is allowed. In other words, no edge incident to the split vertex can be split. In turn, such splitting of edges can increase the domination number. In such a case, we speak of a subdivision number, defined as follows.

The domination subdivision number (or briefly: subdivision number) of a graph G, denoted by $\xi(G)$ or $sd_\gamma(G)$, equals the smallest number of edges that must be divided to create a graph G' for which $\gamma(G') > \gamma(G)$.

Given the above definition, we must assume that finding the dominating subdivision number defined above is only possible for graphs that have three or more vertices (Haynes et al., 2000).

Domination in graphs has its edge counterpart, the definition of which is as follows. The set $X \subseteq E(G)$ is called an *edge-dominating set* of a graph G if every edge that does not belong to the set X is adjacent to some edge in the set X. At the same time, the size of the smallest edge-dominating set of a graph G is called the *edge-domination number* of this graph and is denoted by $\gamma'(G)$.

One of the applications of this type of domination may be the search for critical sets of edges, which can be used in the algorithm for identifying the reliability structures of CNI, which are discussed in Chapter 2. An extension of this application is the use of the concept of a weighted edge domination number, the definition of which is given below, consistent with, among others (Topp, 1993).

The *weighted edge-domination number* $\gamma'_w(G)$ of an edge-weighted graph (G, w_e) is defined as the smallest weight $w(X) = \sum_{e \in X} w_e(e)$ of all edges belonging to the dominating set $X \subseteq E(G)$.

Algorithm 3.4 presents a way to find the minimal edge dominating set. It uses the concept of line graphs defined in Section 2.1.1.

Algorithm 3.4: Minimal-Edge Dominating Set

---

1. Input: $G = (V, E)$
2. Output: minimal edge-dominating set of graph G;
3. define line graphs $L(G)$ of graph G;
4. use algorithm 3.1 for graph $L(G)$;
5. return $D_{L(G)}$
6. $D_E = D_{L(G)}$
7. return D_E.

3.3 ANALYSIS OF CNI SAFETY, RELIABILITY, AND VULNERABILITY

This section begins by introducing general and structural metrics for analysing the CNIs represented by a graph. Some of these metrics are explained in Section 2.1.1. The general measures are distinguished:

Degree centrality – the number of direct connections a vertex has to other vertices in the network.
Closeness centrality – the average distance of the considered vertex to all other vertices in the graph.
Betweenness centrality – the number of shortest paths passing through the vertex.
Edge betweenness – the number of shortest paths passing through an edge.

Graph structural measures characterise the entire graph/network, we distinguish following metrics:

Average degree of a graph vertex
Graph density determines how many connections exist between vertices compared to how many connections are possible between vertices.
Graph diameter measures the length (in terms of edges) of the longest path between any two connected vertices (this is the longest distance among the shortest paths).
Clustering coefficient reflects the degree to which vertices in a graph tend to cluster together.
Mean distance is the average length of the shortest path.

These measures are the base for specific indicators describing transport networks, both at the overall network level and at the node level proposed by Kansky (Rodrigue et al., 2017). Their popularity is related to the fact that they provide:

- a way to enable the expression of relationships between their values and the network structure,
- the ability to compare transport networks at a specific point in time,
- the possibility of comparing the development of transport networks at different points in time.

When we consider the level of the entire network, we take into account the following selected indicators proposed by Kansky(Rodrigue et al., 2017):

- cost (C) expresses the total length of the network measured in actual distances, given by the following formula:

$$C = \sum_{i,j \in V(G)} a_{ij} l_{ij} \tag{3.1}$$

where a_{ij} indicates the existence (value 1) or absence (value 0) of a connection between vertices i and j, and l_{ij} expresses the actual length of the route between vertices i and j;

- detour index is the ratio of the shortest distance $DI(S)$ to the actual $DI(T)$;
- network density determines the territorial occupation of the transport network expressed as the ratio of the length of connections per square kilometer of the area;
- the index Π expresses the ratio of the total length of the graph to the distance along its diameter;
- the index η expresses the average connection length of nodes in the transport network;
- the index Θ is a measure of the node function, expressing the average traffic intensity at a given node in the transport network;
- index β – a measure of connectivity in the graph expressed as the ratio of the number of edges to the number of vertices;
- index α – a measure of connectivity in the graph, which determines the number of cycles compared to the maximum number of cycles in the transport network, described by the formulas:

  for planar graphs

$$\alpha = \frac{u}{2v-5} \tag{3.2}$$

  for non-planar graphs

$$\alpha = \frac{2(e-v)}{v^2-3v+2} \tag{3.3}$$

- index Γ is a measure of connectivity in a graph that expresses the relationship between the number of existing connections and the number of potentially possible ones and is defined by the formulas:

  for planar graphs

$$\Gamma = \frac{e}{3(v-2)} \tag{3.4}$$

  for non-planar graphs

$$\Gamma = \frac{2e}{v^2-v}. \tag{3.5}$$

At the node level, the following metrics can be used (Rodrigue et al., 2017):

- the degree of a node (vertex) is the number of all its neighbours;
- the Koening number (also known as acentricity) – a distance measure based on the number of edges needed to reach the most distant vertex in the graph, expressed by the following formula:

$$e(x) = \max_{y \in X} d(x,y) \tag{3.6}$$

- the Shimbel index (also known as Shimbel distance, vertex reachability) – a reachability measure expressed as the sum of the lengths of all shortest paths connecting all other vertices;

- the betweenness centrality of vertex i ($C_B(i)$) is an accessibility measure expressed as the number of shortest paths passing through vertex i, described by the following formula:

$$C_B(i) = \sum_{l \in V} \sum_{k \neq l \in V} \frac{g_{lk}(i)}{g_{lk}}, \tag{3.7}$$

where $g_{lk}(i)$ is the number of shortest paths passing through vertex i, and g_{lk} is the number of all shortest paths from vertex l to k.

As you can easily see, the advantage of using graphs to describe and model transportation networks is the ability to analyse them at two levels of detail. The first level of detail refers to defining parameters or indicators for the entire network, treating nodes and arcs as a whole. The second level of detail is defined for the nodes themselves, as mentioned above.

3.3.1 CRITICAL NETWORK INFRASTRUCTURE VULNERABILITY ANALYSIS

The application of domination concepts to solving localisation, distance, and representative set problems does not exhaust the possibilities for their other applications. It should be recalled that the use of dominance concepts in graphs allows for:

- minimising the distances to critical points for citizens, given a known number of citizens;
- minimising the number of base stations required to serve all users;
- selecting the minimum number of representatives necessary for comprehensive monitoring of communication networks.

The key to further proposals for applying domination concepts to solving CNI problems is to define the missing parameters or indicators appropriately. These should be a natural extension of those currently in use. Based on this assumption, they can also be divided into specific groups.

When using connected simple graphs to describe a CNI, the smallest dominating sets can indicate the locations of the most essential network nodes, from the perspective of traffic/data/service flow distribution. Thus, the domination (vertex) number is a parameter defining the number of critical nodes in the CNI. Meanwhile, the edge domination number determines the size of the set of critical edges in this network. Nowadays, knowledge of both network characteristics is desirable from the perspectives of traffic engineering and crisis management, as well as when designing reliable and disruption-tolerant networks (e.g., energy, telecommunication, transportation).

When vertex-weighted graphs are used to describe a CNI, the weighted domination number extends the concept to the equivalent in simple graphs. It can provide information about network bottlenecks, for example, due to the short duration of a green light in a specific travel direction or the capacity of the computer network junction. A minimal set of weighted vertices, designated as critical, can provide the basis for determining the so-called "green wave" in transportation networks, which is necessary, for example, for emergency services.

On the other hand, knowledge of the minimal set of vertices provides information about which urban road intersections these services should avoid when planning routes. In such a case, when a certain number of connections between individual nodes (intersections) fail, leading to an increase in the weighted domination number, an undesirable phenomenon occurs, as the power of the set of critical vertices increases. These vertices thus indicate an imperfection in the network. In this case, the weighted equivalent of the bondage-connected number in the graph, as defined formally in this chapter, is a parameter that defines weighted structural vulnerability. This parameter is measured by the number of edges whose removal increases the weighted domination number.

So far, we have discussed the effects of edge removal. However, transportation networks, especially those that underpin the operations of transportation or logistics companies with a specific customer portfolio, may experience growth associated with increasing company shares in the transportation or logistics market. Similarly, we can apply this approach to telecommunications network or energy network. The parameter that determines the sensitivity of such a network to an increase in the number of customers relative to the number of services (located at a node) is the graph subdivision number. On the other hand, it also determines how resilient the selected CNI is to changes in the number of customers.

However, let's consider the bondage connected number of the connected graph, as defined by the author for simple graphs. We obtain a parameter that describes the structural (topological) vulnerability of this network, specifying the number of significant edges whose removal necessitates a change in the network's traffic flow distribution. This involves adding a new set of critical nodes to the existing set, whose task is to absorb a portion of the traffic flow from the network fragment where the edges are removed (no connection). This parameter can be used in online analyses or security modelling of ad-hoc networks to maintain service continuity for generated trips. In specific cases, the search for these significant edges can begin with the set of edges that make up the smallest edge-dominating set mentioned above.

The values of the parameters defining the vulnerability of a CNI should be interpreted as follows. For structural and weighted structural vulnerability, the lower the value of the bondage connected number and the weighted connected bondage number, the more vulnerable the network is. Or in other words, less resilient, the CNI is to service disruptions at its arcs and nodes. A similar relationship applies to the vulnerability of a CNI to an increase in the number of customers; the lower the value of the cohesive linkage number, the greater the network's vulnerability.

Vulnerability testing can be performed at two levels of detail for the graph model representing the CNI under study. Assuming a simple, undirected graph as the model, we analyse the overall vulnerability of its topological structure to various types of disruptions. However, when modelling the CNI using a vertex-weighted or edge-weighted graph, the level of detail increases.

When examining the vulnerability of CNIs, represented by a simple, undirected, and connected graph, it is interesting to identify which edges may cause service problems in this network. For this purpose, it was assumed that the smallest dominating set, or possibly the smallest independent domination set, represents the most important and also the most heavily loaded nodes of the CNI.

As mentioned in Chapter 2, in the work (Fink et al., 1990) on the graph bondage number, this number is treated as a parameter defining the vulnerability of ICT networks. Taking these results into account, the author in another work (Guze, 2019a) proposed new parameters – the connected bondage number and the bondage connected number. Their definitions and preliminary interpretations for CNIs are provided above in this chapter.

These parameters allow for determining the vulnerability of CNI while maintaining their reliability, as defined in Chapter 2. At the same time, established sets of edges whose removal (inefficiency) causes an increase in the corresponding domination numbers are critical sets of a given CNI. Therefore, in everyday use, they require special attention from specialists responsible for the smooth operation of the given network structure (i.e., transportation, energy, telecommunication).

It should also be noted that the graph's bondage numbers indicate a strong relationship between the incapacity of CNI arcs (generally, edges in the graph) and the increase in the load on network nodes. This relationship is expressed by the rise in the domination numbers by at least one.

To illustrate how the new parameters defining the structural vulnerability of a CNI can be used, we consider the following example.

Example 3.1: Transportation Network Analysis

A city area can be divided into traffic regions (Rodrigue et al., 2017), which aids in the analysis and management of traffic flow. The number of these regions depends on the details of the analysis provided. To illustrate this problem, consider an example urban transport network with eight distinct traffic regions, structured as shown in Figure 3.1.

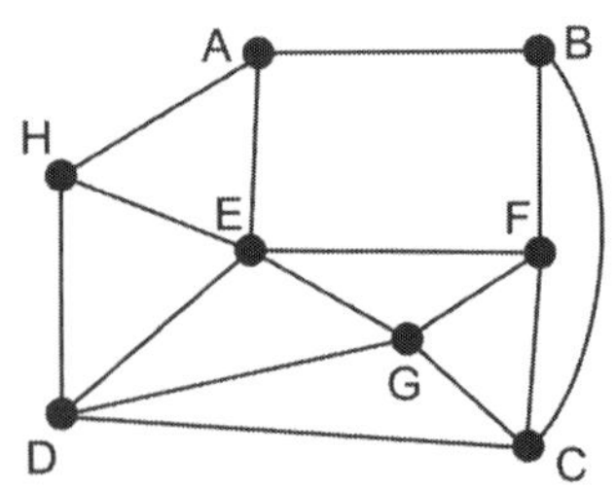

Figure 3.1 A fragment of the urban transport network with eight nodes

In the above graph, we have a set of eight nodes A, B, C, D, E, F, G, H with degree sequence 3, 3, 4, 4, 5, 4, 4, 3. The minimum degree of the graph *G*is 3, and it applies to the vertices A, B, H, while the maximum degree of the graph applies only to the vertex E.

We further assume that the minimum dominating set represents the most essential nodes from the perspective of traffic flow management. In this case, we can formulate the problem of how many faulty connections in the network will necessitate increasing the number of nodes receiving special attention from traffic engineers. In other words, the question is: how vulnerable is the transport network to events blocking traffic on any network arc?

To solve this problem, we can use the domination number and the concept of the number of connected components in a graph. This can be considered in two variants:

- Case 1 { dynamic. After removing arcs in the graph, a new minimum dominating set must be found.
- Case 2 { static. After removing edges, the new minimum dominating set is an extension of the initially found one.

In Figure 3.1, the minimum dominating set of the considered network is defined by the vertices C, E, which were found using Algorithm 3.3. Hence, the domination number of this network is 2. The determined minimum dominating set indicates the nodes that are important in managing the traffic flow of this network.

Now, we can consider the vulnerability of this network in general. The result of finding the bondage connected number for the considered transportation network is given by the edges (C,D), (E,F), (D,E), (D,G), (H,E). The cardinality of the minimum dominating set of the graph increases only after removing five edges (C, D), (D, E), (E, H), (D, G), (E, F), which means that the bondage number of this graph $b_c(G) = 5$. Then the vertices A, G, H are elements of the new minimum dominating set, i.e., D={A, G, H}. These nodes of the urban transport network are currently crucial for traffic flow management.

Let us now consider the distinguished situation in Case 2. We assume the vertices E and C constitute the minimal dominating set. An additional assumption is that this set is fixed (it can only be extended). This means that in the case of examining structural vulnerability using the parameter defined by the bondage number, we extend this minimal dominating set, rather than searching for a new one. It is enough to remove one edge (between nodes B and C) to increase the domination number of graph G. This means that after removing edges based on Algorithm 3.3, vertex F must be added to the dominating set, i.e., the extended minimal domination set is E, C, F.

This approach is more straightforward than the one presented in Case 1, because the traffic flow management system requires less time to respond to changes in the transport network when critical nodes are established. However, the results from the example show that in Case 2, the transport network under consideration is more vulnerable, meaning more susceptible to traffic flow disruptions.

When analysing the above example, it is essential to note that the vulnerability of the transport network is inversely proportional to the value of the bondage connected number in the graph.

Therefore, the failure of a smaller number of edges, leading to an increase in the domination number, indicates greater vulnerability or reduced resilience of the transport network under consideration.

Next, we propose the weighted vulnerability of CNI. It is based on the definition of the weighted bondage connected number in the graph G introduced above. The algorithm for finding the weighted domination number, proposed by the author, whose steps are presented below, facilitates the analysis of transport networks using this parameter.

Example 3.2: Vertex-Weighted Transportation Network Analysis

As an example of the proposed weighted vulnerability methods, consider an intelligent transportation system operating in a selected part of an urban transportation network, which is described by a vertex-weighted graph, as shown in Figure 3.2.

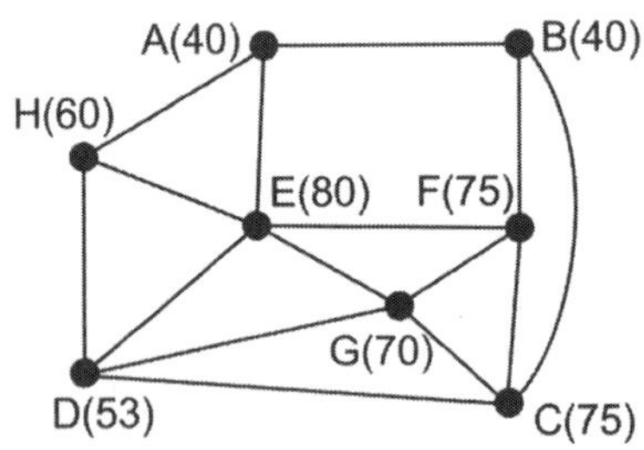

Figure 3.2 A fragment of the vertex-weighted urban transport network with eight nodes

We assume that the weighted dominating set represents the most critical intersections along with the shortest travel time through each intersection. In other words, the weights assigned to the network vertices determine the duration of the red light. Then, the weighted coherence constraint number, $wbc(G)$, provides information about the number of blocked network arcs that increases the number of intersections requiring more attention and necessitates a change in the red light duration. As in the previous section, the solution can be applied in two cases:

- Case 1 dynamic – after removing arcs in the graph, a new minimum dominating set must be found.
- Case 2 static – after removing edges, the new minimum dominating set is an extension of the initially found one.

With respect to the defined transport network, the minimum weighted dominating set found by Algorithm 3.5 is determined by the $D_w = B, D$. At the same time, the weighted dominating number is 93, i.e., $\gamma_w(G) = 93$. In solving the static problem (case 2), we assume that this set forms the basis of the new minimum weighted dominating set.

Then, using the weighted connected bondage number as a parameter of weighted structural vulnerability, we can conclude that in Case 1, removing the edge (A, B) is enough to increase the weighted dominant number to 133. However, let's consider the dynamic case in which, after the edge removal operation, we search again for the lightest dominant set. We can state that the value of the

Algorithm 3.5: Weighted Domination Number

1. Input: graph (G, w_v) with a vertex weighting function.
2. Output: weighted domination number of graph G $\gamma_w(G)$.
3. For all for $i = 1$ to $|V|$, do
   a. find the dominating set of G starting from $v_i \in V$ using Algorithm 6.1 or 6.2 and write it as the dominating set $D_i(v_i)$;
   b. find the total weight $W_i(v_i) = \sum_{v_i \in D_i} w_i(v_i)$ of the dominating sets $D_i(v_i)$;
4. End.
5. Find $\min\{W_i(v_i) : v_i \in D_i(v_i), i = 1, \ldots, |V|\} \to \gamma_w(G)$.
6. Return $\gamma_w(G)$.
7. End.

weighted structural vulnerability parameter is 1, and, similarly to the first case, it is enough to remove the edge (A, B) for the weighted domination number to increase to 125 instead of the nominally found 93.

Next concept, which can help us in vulnerability or resilience analysis, is the algorithm introduced as the basis for further theoretical and practical results is a deterministic discrete-time model of fire spread on a graph G. Hartnell (1995) introduces this concept and calls *firefighter problem*. This model of fire spread consider how firefighters can act to stop a fire outbreak. An outbreak of fire starts at a set of root vertices of G at time $t = 0$. In response, firefighters defend f vertices at time $t = 1$ (Hartnell, 1995). The pseudocode of proposed algorithms for finding solution of this problem is presented as Algorithm 3.6.

Algorithm 3.6: Firefighting Problem (or Virus Control)

1. Input: graph $G = (V, E)$.
2. Output: the set of vertices burning and saved.
3. A fire breaks out at a vertex of a graph.
4. The firefighter (or defender) then chooses any vertex not yet on fire (or affected by the virus) to protect.
5. The fire and firefighter alternate moves on the graph due to the fire can no longer spread.
6. The firefighter has chosen a vertex; it is protected or safe from future fire moves.
7. After the firefighter's step, the fire spreads to all vertices adjacent to the ones on fire, except those protected.

According to the firefighting problem, we call a vertex burning or burned if it is labelled as burning. A vertex labelled as defended is called defended. Since the label of a vertex does not change, once a vertex is burning or defended, it remains so. If a vertex never burns, it is called saved. The damage is the total number of vertices with label "burning" at the end of the process. For a finite graph with n vertices, the number of vertices saved plus the damage equals n. The notation $mvs(G,F;d)$ represents the maximum number of vertices that can be saved in G when fires break out at vertices in F, considering all possible ways of defending vertices using d firefighters. The notation $mvb(G,F;d)$ represents the maximum number of vertices that can be burned in G when fires break out at vertices in F, considering all possible ways of defending vertices using d firefighters. Accoridng to this concept, we introduce two new parameters for vulnerability and resilience analysis. A maximal number of saved vertices of graph G, $msv(G)$ defines the power of a maximal set of saved vertices. It can be interpreter as the CNI level of resilience.

Furthermore, the maximal number of burned vertices in graph G, $mbv(G) = n - msv(G)$ can describe the CNI level of vulnerability.

3.3.2 GAME THEORY APPROACH TO ANALYSIS CRITICAL NETWORK INFRASTRUCTURE

Game theory, the foundations of which are presented in Section 2.1.2, is intended in this book to provide decision-making models (both operational and functional) for CNIs. It can be used to support decision-making under conditions of uncertainty – that is, when information is uncertain or incomplete. Games with nature are used for this purpose. This subchapter also proposes a specific model of games with nature, where the states of nature are the individual elements of a SWOT (strengths, weaknesses, opportunities, and threats) analysis. This allows for a two-stage decision-making model (see 3.7. In the first step, we conduct a SWOT analysis, but with the caveat that it must be expressed both quantitatively and qualitatively. Then, based on this, in the second stage, we construct a payoff matrix for specific states of nature and the considered scenarios. Then, we use the known criterion described in Section 2.1.2 to solve this game. Such a model will likely help select strategies to improve the security, efficiency, and effectiveness of actions that mitigate, strengthen, and restore the whole usability of CNIs.

We recognise that SWOT analysis is a valuable tool in strategic planning. In the case of CNIs, it involves selecting and allocating resources to restore the whole service operation following an unexpected event. The method enables us to connect internal resources and the external environment, allowing us to utilise opportunities more effectively and manage threats better. A SWOT analysis employs a straightforward classification scheme that enables the categorisation of all factors influencing the current and future position of the CNI. We can distinguish four categories of factors: external positive opportunities, external damaging threats, internal positive strengths, and internal opposing weaknesses as part of this analysis. In engineering applications, SWOT analysis must be combined with expert knowledge. This can be obtained using Likert-scale surveys. This knowledge enables the numerical representation of individual components of the SWOT analysis. Using these weights, we can plot the graph with strategic points, which symbolise a vital position based on a mathematical comparison of the sum of points on the appropriate 0-T and S-W axes. To do this, we subtract these values, subtracting less from the more significant. The intersection of the lines perpendicular to the axes that pass through the mentioned points determines the strategic position.

The intersection of the lines perpendicular to the axes that pass through the mentioned points determines the strategic position. In the economy, the location of the point represents the company's strategic position. We can point out the following positions:

The *maxi-maxi position* gives privilege in the market – it is a preponderance of strengths over weaknesses and opportunities over threats.
The *maxi-mini position* means that the company's strengths outweigh its weaknesses, but the environmental threats outweigh the opportunities. The organisation's main effort should be to mitigate the impact of threats and capitalise on remaining market opportunities based on its strengths.
The *mini-maxi position* means opportunities outweigh threats, but unfortunately, the company's weaknesses outweigh its strengths. The company's management must focus on improving the company's condition in inefficient (non-competitive) areas.
The *mini-mini position* represents an inefficient enterprise, where weaknesses outweigh strengths and threats outweigh opportunities. The company functions in an environment without opportunities for beneficial action.

In terms of the scenario for strategic planning to mitigate or strengthen the resilience of the CNI, the above items may indicate varying levels of effectiveness. However, after determining the position of a given scenario for mitigating a specific threat, the next step is to select the best one. This raises the issue of uncertainty regarding the ultimate effectiveness. To choose the best scenario, we can utilise game theory in conjunction with nature (see Section 2.1.2). To do this, we construct a payoff matrix based on the states of nature defined by S, W, O, and T. Thus, we get the payoff matrix in the form presented in Table 3.1

Table 3.1
General Schema of a SWOT Game

	Nature Contingencies			
Strategies	**Strengths**	**Weaknesses**	**Opportunities**	**Threats**
Strategy 1	Payoff's1	-Payoff'w1	Payoff'o1	-Payoff't1
Strategy 2	Payoff's2	-Payoff'w2	Payoff'o2	-Payoff't2
Strategy 3	Payoff's3	-Payoff'w3	Payoff'o3	-Payoff't3

In the application of the theory of games with nature to SWOT analysis, negative values in the state of nature for weaknesses and threats should be considered, as the result would be falsified if we assumed that the highest score obtained is the decisive value for the choice. Particularly, it is true in the case of the Laplace-Bayes criterion. Without considering this relationship, the decision-maker choosing the highest score could select a decision that generates weaknesses and threats. We build a SWOT matrix for each strategy to consider. In this place, we propose a general decision-support Algorithm 3.7, based on Strzelczyk and Guze (2025).

Algorithm 3.7: SWOT Game with Nature

1. Input: Decision problem.
2. Output: optimal or quasi-optimal strategy.
3. Define the decision problem.
4. Make the SWOT analysis.
5. Built game with nature, where S, W, O, and T are the nature states.
6. Resolve the game according to known decision criteria described in Subsection 2.2.
7. Find the EMV, EOL, EVPI.
8. Give the decision about the strategy based on steps 2–5.

The EMV, EOL, and EVPI are defined as follows. *Expected opportunity loss (EOL)* refers to the evaluation of various alternatives. The losses from making a specific decision are assessed, considering all possible outcomes and probabilities. This parameter compares alternatives and chooses the one that minimises the expected loss of opportunity according to formula:

$$EOL(a_i) = \sum_{j=1}^{m} s_{ij} \; x \; p_j \tag{3.8}$$

where

a_i – i-th strategy (i-th row of payoff matrix);
p_j – j-th nature's state (j-th column of payoff matrix);
s_{ij} – payoff value for strategy a_i with probability of nature's state p_j.

Expected monetary value (EMV) assesses the risk associated with a project. This method estimates the financial value associated with the potential risks. It allows you to analyse different scenarios and assign them appropriate expected values to assess the impact of risk on the project's value. This method is based on the probability of a given risk scenario occurring. The EMV for choosing the i-th strategy (a_i) for the j-th nature's state probability (p_j) is as follows (Tulsian and Pandey, 2016):

$$EMV(a_i) = \sum_{j=1}^{m} W_{ij} \; x \; P_j \tag{3.9}$$

where

a_i – i-th strategy (i-th row of payoff matrix);
p_j – j-th nature's state (j-th column of payoff matrix);
W_{ij} – payoff value for strategy a_i with probability of nature's state p_j.

Finally, the *expected value of perfect information (EVPI)* measures the theoretical maximum value that can be achieved by having complete and accurate information before making a decision, which can influence the decision and reduce risk. The EVPI value is the difference between the expected value of the outcome with complete(perfect) information (EVWPI) and the expected value without this information (EVWOPI), which is equal to the EMV. EVWPI is calculated according to the following formula (Jackson et al., 2024):

$$EVWPI\,(a_i) = \sum_{j=1}^{m} \max_i \,(w_{ij})\; x\; P_j\,. \tag{3.10}$$

Thus, the formula for EVPI is as follows:

$$EVPI = EVWPI - EMV. \tag{3.11}$$

Defender-attacker game model

Cybersecurity is becoming increasingly important year by year, as most public, financial, and educational services are moving to the virtual world. Therefore, we need mathematical and decision-making models that allow us to analyse various scenarios of hacker attacks on individual elements of network infrastructure.

In this respect, defender-attacker games come to our aid. According to the theory presented in Chapter 2, this is formally a non-cooperative two-player zero-sum game represented by $SG = \langle PL, A, U\rangle$, where $PL = \{A, D\}$ represents two players Player A is a malicious node/attacker and the other player D is a defender. $A = \{a_r, d_r | r \in \{0,1\}\}$ is the strategy space, and U is the utility for each player. In the simple game, the attacker and the defender can use one of the two levels of the available strategies during the game. For the attacker, level one means a low intensity of attack, denoted by $a1$="Attack-1," level two is a high intensity of attack, denoted by $a2$="Attack-2." Correspondingly, level one for the defender means that they decide to implement low level of defense, denoted by $d1$ ="Defend-1," and level two is a high intensity of defense, denoted by $d1$="Defend." Both players choose their strategies simultaneously without any collaboration, assuming common knowledge about the game.

Furthermore, we assume that the value of the protected assets by the defender D is equal to ω_n, where $\omega_n > 0$ and $\omega \in \{1,2\}$. Then, ω_1 is the value of assets compromised by the Attack-1 strategy deployed by the attacker successfully; ω_2 is the value of assets compromised by the Attack-2 strategy deployed by the attacker successfully. We also assume that the gain of one player is equal to the loss of the other, because it is a zero-sum game. Thus, ω_n is the gain by the attacker if his strategy Attack-n is successful and $-\omega_n$ denotes the loss/damage by the defender. The value of this loss by the defender refers to the degree/amount of damage, such as wasting energy, number of compromised/disabled nodes, loss of data integrity, lost connection to services, etc.

On the other hand, the attacker/defender also needs to pay a certain cost to implement their attack/defence strategies. For the attacker, we denote the cost of attack as c_{a_n}, where $n \in \{1,2\}$: c_{a_1} is the cost to deploy Attack-1 strategy, and c_{a_2} is the cost to deploy Attack-2 strategy. Likewise, for the defender, we denote the cost of defence as c_{d_n} where $n \in \{1,2\}$: c_{d_1} is the cost to deploy the Defend-1 strategy, and c_{d_2} is the cost to deploy the Defend-2 strategy.
Based on this assumptions, we draw the payoff matrix presented as Table 3.2.

Table 3.2
Payoff Matrix for a Simple Defender-Attacker Model Game – Single Step

		Defender (D)	
	Strategies	d_1	d_2
Attacker (A)	a_1	$c_{d_1} - c_{a_1}, c_{a_1} - c_{d_1}$	$c_{d_2} - c_{a_1}, c_{a_1} - c_{d_2}$
	a_2	$\omega_2 + c_{d_1} - c_{a_2}, c_{a_2} - c_{d_1} - \omega_2$	$c_{d_2} - c_{a_2}, c_{a_2} - c_{d_2}$

It is obvious that $c_{d_1} < c_{d_2}$ and $c_{a_1} < c_{a_2}$, and $\omega_n > c_{a_n}, c_{d_n}$ $(n = \{1,2\})$.

We can extend this approach to use the firefighting problem defined in Section 3.2. Then we assume that "fire" is the attacker, and "fireman" is the defender. For each iteration of the firefighter algorithm, we can write defender-attacker game matrices. For example, we can assume the defence costs are the number of firefighters required to stop the spread of fire in the graph. Combining the graph-theoretic algorithm with the game model discussed here provides a practical and comprehensive tool for analysing CNIs from the perspective of their resilience to any disruptive events. Of course, time must be devoted to calibrating such models – in other words, training them. However, after appropriately training the analytical model with the training set, we also gain the ability to use neural networks to conduct this type of analysis.

3.4 CASE STUDIES

This section describes three case studies. First is about US pipeline network analysis. Second case analyse Saudi Arabia pipeline network. Third case is about Baltic Port Network (BPN).

3.4.1 VULNERABILITY ANALYSIS OF USA PIPELINES

Let us consider the graph model of US pipelines from Section 2.4. Now, we analyse a structural vulnerability of this network. Figure 3.3 presents the minimal dominating set – rounded vertices constitutes him. There are vertices in this set, so the domination number equals to 19. These 19 nodes need to be secure – they are critical for this pipeline.

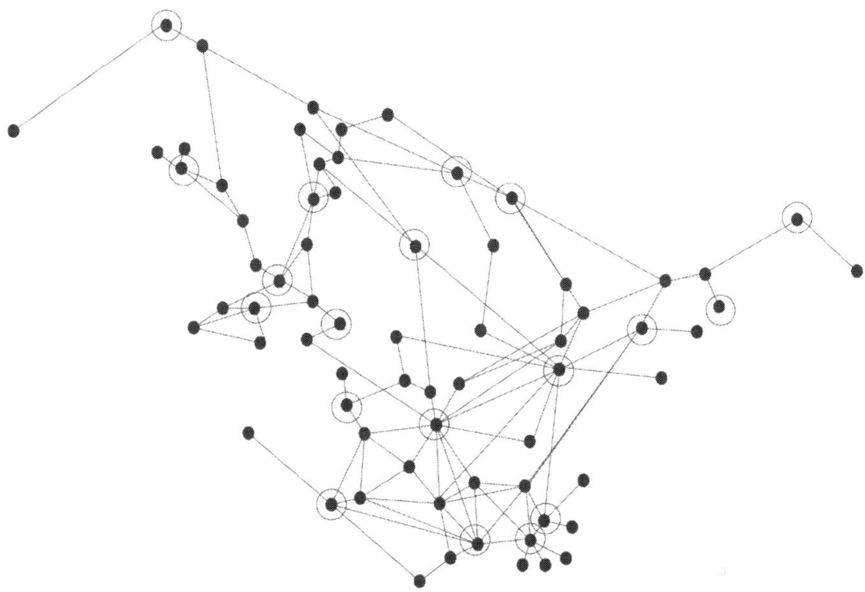

Figure 3.3 Minimal dominating set of US pipeline network

We want to get the answer for question: How vulnerable is this network to disruptions related to the operation of connections between individual nodes?
To find the solution we have to find the bondage connected number. Similarly to example discussed in Section 3.2, we can consider two cases:

Case 1 (dynamic) – after removing arcs in the graph, a new minimum dominating set must be found.
Case 2 – (static) – after removing edges, the new minimum dominating set is an extension of the initially found one.

Considering Case 1, it is clear to find that it is enough to remove fourth edges to increase domination number (new dominating set). It means that the network will need be secured for these five links between junctions. In Case 2, domination number increases also after removing one edge. It only takes a single disruption to increase the domination number.

3.4.2 VULNERABILITY ANALYSIS OF SAUDI ARABIA PIPELINES

Let us consider the graph model of Saudi Arabia pipelines from Section 2.4. Now, we analyse a structural vulnerability of this network. Figure 3.4 presents the minimal dominating set – rounded vertices constitutes him. There are 10 vertices in this set, so the domination number equals to 10.

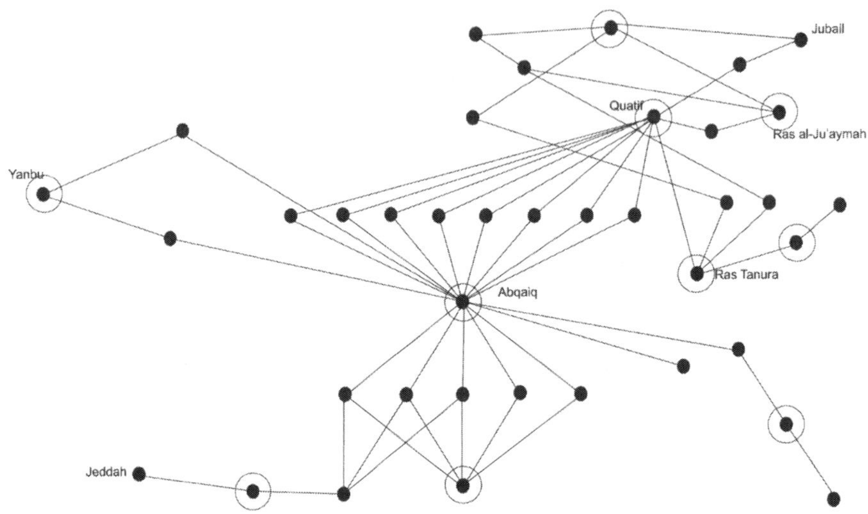

Figure 3.4 Minimal dominating set of Saudi Arabia pipeline network

We want to get the answer for question: How vulnerable is this network to disruptions related to the operation of connections between individual nodes?
To find the solution we have to find the bondage connected number. Similarly to example discussed in Section 3.2, we can consider two cases:

> Case 1 – (dynamic) – after removing arcs in the graph, a new minimum dominating set must be found.
> Case 2 – (static) – after removing edges, the new minimum dominating set is an extension of the initially found one.

Considering Case 1, it is clear to find that it is enough to remove one edge to increase domination number (new dominating set). In Case 2, domination number increases also after removing one edge.

It only takes a single disruption to increase the dominance number. Depending on whether the node corresponds to an oil field, a port terminal, or a major refinery, the consequences for crude oil exports will vary. However, the most tremendous consequences would be the disruption of the connection to one of the ports on Saudi Arabia's eastern coast.

Now, we try to show how to apply the Algorithm 3.6. We can assume that the ports Ras al-Ju'aymah, Jubail, Yanbu, and Ras Tanura have the protection. Similarly, we can assume the same about the largest oil field at Abquaiq. Thus we know that five vertices in oil pipeline network are protected. So, the "fire" can start at Quatif. Then "fireman" goes to Ras Tanura. Following the algorithm, all neighbours of Quatif are with label fire, but not Ras Tanura.

At this point, we can already conclude that damage to the Quatif node would result in the disconnection of half of Saudi Arabia's oil pipeline system (ca.). This would reduce the country's export capacity, which would invariably lead to higher oil prices. A failure that would take the Abqaiq oil field out of production would have even greater consequences for this network, the country, and the world. At this point, Saudi Arabia's production capacity would drop drastically.

Therefore, if we return to the smallest dominating set (one of the possibilities) shown in the figure, we can see why this set indicates nodes of this CNI requiring special attention.

3.4.3 BALTIC PORTS NETWORK

The Baltic Sea has a surface area of approximately 415,000 km^2, a volume of roughly 22,000 km^3, and a developed coastline of approximately 22,000 km. The Baltic Sea stretches from north to south for almost 1,300 km and is half as long from east to west – about 600 km. The narrowest part of the Baltic Sea – the Gulf of Bothnia – is about 100 km wide.

Major ports of the Baltic Sea include Aarhus, Copenhagen–Malmo, Lubeck, Rostock, Tallinn, Riga, Ventspils, Klaipeda, Gdansk, Gdynia, Szczecin–Swinoujscie, Helsinki, Turku, Hamina–Kotka, Gothenburg, Lulea, Stockholm, and Trelleborg.

Based on these ports, we can define the *BPN*, which we describe using a graph where the vertices represent the specified ports and the edges represent existing maritime connections between them. Figure 3.5 presents a graph model representing the BPN. Due to the assumption made throughout the work that we consider connected graphs, the Lulea port is included in Figure 3.5 because there is no maritime connection to the other BPN ports.

To simplify notation, the individual ports in this figure are described by consecutive natural numbers, where: 1 – Hamina–Kotka, 2 – Helsinki, 3 – Turku, 4 – Tallinn, 5 – Stockholm, 6 – Riga, 7 – Ventspils, 8 – Klaipeda, 9 – Gdansk, 10 – Gdynia, 11 – Szczecin–Swinoujscie, 12 – Rostock, 13 – Lubeck, 14 – Aarhus, 15 – Gothenburg, 16 – Copenhagen–Malmo, 17 – Trelleborg.

For the transportation network presented in this way, a structural vulnerability analysis can be performed, and its reliability structure can be determined in the edge, vertex, and general sense based on the research results presented in chapters 2 and 3.

Firstly, using Algorithm 3.3, the smallest dominating set was found, consisting of the vertices 1, 4, 7, 17. It is worth noting that this is also the smallest independent dominating set.

Analysing the sensitivity of the BPN, we can conclude that the bondage connected number of the graph representing it is 7. This value indicates that removing at least

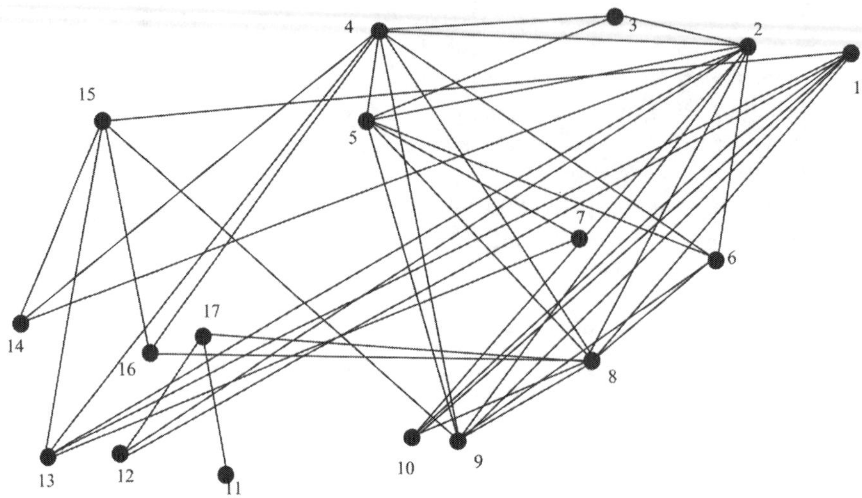

Figure 3.5 Graph model of the BPN

seven edges from the graph will increase the domination number by 1. The edges that need to be removed are (1, 4), (2, 4), (4, 5), (5, 7), (4, 13), (4, 16), (2, 3). This means there are no connections between pairs of ports:

Hamina–Kotka and Tallinn
Helsinki and Tallin
Tallin and Stockholm
Stockholm and Ventspils
Tallin and Lubeck
Tallin and Copenhagen–Malmo
Helsinki and Turku.

However, the identification of reliability structures in both the edge and vertex senses indicates that this is a serial transport network in both cases. In the edge case, there is a bridge in the considered transport network that connects vertices 11 and 17. Considering the reliability structure in the vertex sense, vertex 17 exists in the graph representing the BPN; it is a cutting vertex.

Therefore, based on the results of identifying reliability structures in the vertex and edge senses, the BPN is of the serial-serial type, i.e., in accordance with the previously introduced nomenclature, serial.

4 Optimisation of Critical Network Infrastructures

The chapter begins with an introduction to the basic methods, techniques and algorithms of single- and multi-objective optimisation. This information is supplemented with elements of game theory, including games with nature. The next part of the chapter presents optimisation methods and algorithms based on previously introduced methods of modelling critical network infrastructures (CNIs). In particular, a new approach is shown, considering the description of elements using continuous weight functions. Thus, gradient methods are adapted to optimise CNIs in particular.

4.1 GENERAL APPROACH AND BASIC NOTATIONS

To solve optimisation problems, we can use the following well-known notations. In the case of single-criteria optimisation, we consider the following objective function (Guze S. Kołowrocki, 2017; Zambrano-Martinez et al., 2020; Salas et al., 2007):

$$\begin{aligned} F(x_i) &\rightarrow \min / max \\ c_j(x_i) &\leq 0 \\ c_j(x_i) &\geq 0 \\ x_i \geq 0,\ i,j &= 1,2,\ldots,n \end{aligned} \tag{4.1}$$

where

x_i are decision variables, with $i = 1,2,\ldots,n$;
$F(x_i)$ are objective functions;
$c_j(x_i)$ are constraint functions (upper or lower) of decision variables, with x_i, $i,j = 1,2,\ldots,n$.

In the case of multi-criteria optimisation, we consider the model described by a vector function **f** (or shorter by a vector), which maps m decision variables on the n functions of goal (criteria) with k sets of constraints. It assumes the following form (Guze S. Kołowrocki, 2017; Salas et al., 2007; Venter, 2010):

objective functions

$$\mathbf{y} = \mathbf{f}(\mathbf{x}) = [f_1(\mathbf{x}), f_2(\mathbf{x}), \ldots, f_n(\mathbf{x})] \rightarrow \min\ /\ \max \tag{4.2}$$

constraints

DOI: 10.1201/9781003661450-4

minimisation problem

$$\mathbf{o}(\mathbf{x}) = (o_1(\mathbf{x}), o_2(\mathbf{x}), \ldots, o_k(\mathbf{x})) \geq \mathbf{0} \tag{4.3}$$

maximisation problem

$$\mathbf{o}(\mathbf{x}) = (o_1(\mathbf{x}), o_2(\mathbf{x}), \ldots, o_k(\mathbf{x})) \leq \mathbf{0} \tag{4.4}$$

and under assumption:

$$\begin{aligned} \mathbf{x} &= (x_1, x_2, \ldots, x_m) \in X, \\ \mathbf{y} &= (y_1, y_2, \ldots, y_n) \in Y \end{aligned} \tag{4.5}$$

where

$\mathbf{x}$ is the m-dimensional vector of decision variables;
X is the space of decision variables;
$\mathbf{y}$ is the n-dimensional vector of objective functions (decision criteria);
Y is the space of objective functions (decision criteria);
$\mathbf{o}(\mathbf{x})$ is the vector of decision-variable constraints.

It should be noted that the objective and constraint functions are functions of the decision variables.

Using constraints on the vector of decision variables $\mathbf{o}(\mathbf{x}) \leq \mathbf{0}$ ($\mathbf{o}(\mathbf{x}) \geq \mathbf{0}$), the set of possible solutions to the maximisation (minimisation) problem is described. Then, the set:

$$\mathbf{X}_f = \{\mathbf{x} \in \mathbf{X} | \mathbf{o}(\mathbf{x}) \leq \mathbf{0}\} \tag{4.6}$$

$$(\mathbf{X}_f = \{\mathbf{x} \in \mathbf{X} | \mathbf{o}(\mathbf{x}) \geq \mathbf{0}\}) \tag{4.7}$$

decision vectors $\mathbf{x}$, which are subject to the constraints $\mathbf{o}(\mathbf{x}) \leq \mathbf{0}$ ($\mathbf{o}(\mathbf{x}) \geq \mathbf{0}$), are called collective constraints for the maximisation (minimisation) problem. The question is that its image, i.e., the regions available in the objective function space, formula (Zitzler and Thiele, 1999):

$$\mathbf{Y}_f = \mathbf{f}(\mathbf{X}_f) = \bigcup_{\mathbf{x} \in \mathbf{X}_f} \{\mathbf{f}(\mathbf{x})\} \tag{4.8}$$

In the practice of solving optimisation problems for transport and computer networks, the use of linear programming or bi-level programming methods dominates.

According to Kalashinkov et al. (2015) and Kuang et al. (2013), the linear programming problem is based on the mathematical programming problem that occurs when the objective function and the constraint functions in the decision problem (7.1) are linear functions.

The bi-level programming method, on the other hand, is based on the partitioning of decision variables between two vectors $\mathbf{x}$ and $\mathbf{y}$. Vector $\mathbf{x}$ is chosen as the optimal solution to the second problem parameterised in vector $\mathbf{y}$. This method is therefore classified as hierarchical, as its constraints are partially defined by the second optimisation problem, which in practice is introduced first. The general form of the second-level problem, also called the follower's problem, can be written as the relation (Kalashinkov et al., 2015; Kuang et al., 2013):

$$\min_{x} \{f(x,y) : g(x,y) \leq 0, h(x,y) = 0\} \tag{4.9}$$

where

$f : R^n \times R^m \to R$ – lower-order objective function,
$g : R^n \times R^m \to R^p$, $h : R^n \times R^m \to R^q$ – limiting functions of the form:

$$\begin{aligned} g(x,y) &= [g_1(x,y), g_2(x,y), \ldots, g_p(x,y)]^T \\ h(x,y) &= [h_1(x,y), h_2(x,y), \ldots, h_q(x,y)]^T. \end{aligned} \tag{4.10}$$

Next, we assume that the function $\Psi(y) : R^m \to 2^{R^m}$ is a solution to the above problem for a fixed $y \in R^m$. Additionally, we establish that $x(y)$ is some element of the function $\Psi(y)$ chosen uniquely for all y. This choice is made in such a way that equalities (in particular, nonlinear ones) or limiting inequalities written in the general form:

$$G(x(y),y) \leq 0, H(x(y),y) = 0 \tag{4.11}$$

were satisfied.

Then the objective function is minimised according to the rule:

$$\text{"}\min_{y}\text{"} \{F(x(y),y) : G(x(y),y) \leq 0, H(x(y),y) = 0, x(y) \in \Psi(y)\} \tag{4.12}$$

where

$F : R^n \times R^m \to R$ – higher-order objective function,
$G : R^n \times R^m \to R^k, H : R^n \times R^m \to R^l$ – higher-order vector limiting functions, represented in the general case by the expressions:

$$G(x,y) = [G_1(x,y), G_2(x,y), \ldots, G_k(x,y)]^T \tag{4.13}$$

$$H(x,y) = [H_1(x,y), H_2(x,y), \ldots, H_l(x,y)]^T \tag{4.14}$$

The above problem is called the two-level programming problem or the leader's problem in the literature (Kalashinkov et al., 2015; Kuang et al., 2013).

4.2 PRINCIPLES OF THE PARETO OPTIMALITY

In practice, searching for optimal solutions to multi-criteria problems usually does not lead to exact solutions, but only to approximate ones. Therefore, the approach should be such that the solution set to the multi-criteria optimisation problem consists of all decision vectors for which the corresponding criterion vectors cannot be improved in any dimension without deterioration in another. These vectors are known as Pareto optimal, which is related to the concept of vector domination by another. To facilitate understanding, well-known definitions should be cited (Zitzler and Thiele, 1999).

Let us consider the maximisation (minimisation) problem and consider two decision vectors $\mathbf{a}, \mathbf{b} \in X$:

1. We say that **a** dominates **b** if and only if:

$$\forall i \in \{1,2,\ldots,n\}: \ f_i(\mathbf{a}) > f_i(\mathbf{b}) \wedge \exists j \in \{1,2,...,n\}: \ f_j(\mathbf{a}) > f_j(\mathbf{b}), \tag{4.15}$$

$$(\forall i \in \{1,2,\ldots,n\} : f_i(\mathbf{a}) < f_i(\mathbf{b}) \wedge \exists j \in \{1,2,\ldots,n\} : f_j(\mathbf{a}) < f_j(\mathbf{b}). \tag{4.16}$$

2. We say that **a** weakly dominates **b** if and only if:

$$\forall i \in \{1,2,\ldots,n\}: \ f_i(\mathbf{a}) \geq f_i(\mathbf{b}) \wedge \exists j \in \{1,2,\ldots,n\}: \ f_j(\mathbf{a}) \geq f_j(\mathbf{b}), \tag{4.17}$$

$$(\forall i \in \{1,2,\ldots,n\} : f_i(\mathbf{a}) \leq f_i(\mathbf{b}) \wedge \exists j \in \{1,2,...,n\} : f_j(\mathbf{a}) \leq f_j(\mathbf{b})). \tag{4.18}$$

3. We say that **a** is indifferent to **b** if and only if:

$$\forall i \in \{1,2,\ldots,n\}: \ f_i(\mathbf{a}) not \geq f_i(\mathbf{b}) \wedge f_i(\mathbf{b}) not \geq f_i(\mathbf{a}), \tag{4.19}$$

$$(\forall i \in \{1,2,\ldots,n\} : f_i(\mathbf{a}) not \leq f_i(\mathbf{b}) \wedge f_i(\mathbf{b}) not \leq f_i(\mathbf{a})). \tag{4.20}$$

Now, let $\mathbf{X}_f$ denote the set of feasible solutions. Then, a decision vector $\mathbf{x} \in \mathbf{X}_f$ is called non-dominated in the set $A \subseteq \mathbf{X}_f$ if and only if there is no vector $\mathbf{a} \in A$ that dominates the vector $\mathbf{x}$ (i.e., $\mathbf{a} \succ \mathbf{x}$). In other words, all decision vectors that are not dominated by other decision vectors are called non-dominated. In such a case, Pareto optimality can be defined as follows (Zitzler et al., 1998). A decision vector $\mathbf{x}$ is called Pareto optimal if and only if it is non-dominated in the set $\mathbf{X}_f$.

Hence, when decision vectors are non-dominated across the entire search space of feasible solutions, they are called Pareto optimal and form a Pareto-optimal front. Figure 7.1 shows the set of feasible solutions, indicating the dominated ones (black points) and the non-dominated ones forming the Pareto front (white points). This defines a set of Pareto-optimal options (white points). This allows the decision maker to make trade-offs within this set, rather than considering the full range of each parameter. This means that the shape of the Pareto front indicates the nature of the trade-off between different objective functions, which addresses the problem of finding exact values for vectors of decision variables.

Considering statistical decision theory, the above approach can be compared to an admissible decision rule. This is a rule for making decisions in such a way that no other rule is always "better" than it. Generally speaking, the set of admissible rules for most decision problems is large, sometimes infinite. Therefore, this is not a sufficient criterion for considering a single rule, but it should favour admissible rules. Pareto optimality suggests which variant a decision-maker may think is best.

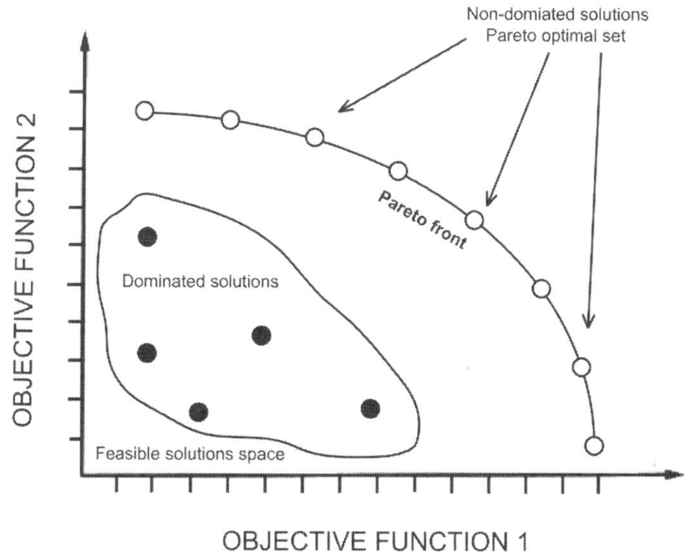

Figure 4.1 An example of a Pareto-optimal front – maximisation problem

At this point, we should also return briefly to game theory. Pareto optimality also plays a significant role here. This is particularly true, though not limited to, economic applications of game theory.

In game theory, Pareto optimality refers to a player's action profile. We say that a strategic game profile is Pareto optimal if no profile provides at least one player with a higher payoff and all others with at least the same payoff. Furthermore, a game profile is Pareto suboptimal if there exists another profile that is better for at least one player and no worse for any (i.e., when it is not Pareto optimal).

The presented approach is used in this book as a basic approach for large-scale optimisation of problems related to CNIs.

4.3 OPTIMISATION METHODS FOR CRITICAL NETWORK INFRASTRUCTURES

In this section, we present the the literature on transportation network optimisation is dominated by problems related to minimising travel time (Holmgren, 2004a; Johnson and McGeoch, 1997; Kalashinkov et al., 2015; Lam and Zhang, 2000; Lawler et al., 1985; Ming-Hua et al., 2012; Tinós, 2015,?) or distance traveled (Guze S. Kołowrocki, 2017; Holmgren, 2004b; Johnson and McGeoch, 1997; Kalashinkov et al., 2015; Lawler et al., 1985; Nelder and Mead, 1965; Nesetril et al., 2001; Zhan and Noon, 1998). Another group of problems concerns maximising the flow of goods or passengers in transport networks (Ford and Fulkerson, 1956; Guze S. Kołowrocki, 2017; Kalashinkov et al., 2015; Lawler et al., 1985; Shariat Mohaymany and Babaei,

2013). An essential part of the research is also devoted to the issue of optimal design of transport networks so that they meet the above criteria and, additionally, are resistant to various natural factors and those resulting from deliberate, harmful human activity.

Therefore, the described transportation network optimisation problems can be broadly grouped as follows:

- traffic distribution and management within the network to ensure smooth traffic flow,
- planning and designing transportation networks to minimise the time and costs associated with operating transport services,
- crisis management in transportation networks with particular emphasis on natural factors and harmful human activity – network vulnerability and network resilience.

In case of telecommunication networks the optimisation problems are as follows:

- designing networks
- network reliability and survivability with connectivity restrictions
- Quality of Service
- simultaneous network synthesis: loading capacity
- non-simultaneous network synthesis: back-up capacity
- accessibility to a network
- backbone network
- resource allocation problem
- routing problem

When we consider the optimisation problem in the energy networks, we actually have the following exemplary issues defined:

- electricity generation, storage, and distribution
- optimising energy flows and connections between energy networks
- investment planning and operation management of distributed electricity supply systems
- design, operation and sizing of distributed energy supply systems.

4.3.1 DISCRETE APPROACH TO MULTI-CRITERIA OPTIMISATION

This section begins with a discussion of CNI optimisation through its reducibility to the 0–1 knapsack problem. The literature on multi-objective discrete optimisation is dominated by solutions based on linear and bi-level programming (e.g., transportation networks (Kalashinkov et al., 2015; Kuang et al., 2013)). As already mentioned, the latter method in particular has certain limitations and is very difficult to implement. Therefore, at least one stage of this method requires the use of evolutionary algorithms or other methods based on approximate solutions. Among these methods, the following are worth mentioning (Zitzler and Thiele, 1999):

Strength Pareto Evolutionary Algorithm – SPEA
Vector Evaluated Genetic Algorithm – VEGA
Hajela and Lin's Weighting-based Genetic Algorithm – HLGA
Niched Pareto Genetic Algorithm – NPGA

The general theory of multi-objective optimisation also includes analytical methods that can return results in a deterministic or nondeterministic manner. Some of them are (Guze, 2015):

weighted objective methods
hierarchical optimisation method
trade-off method
global criterion method
method of distance functions
min–max methods
goal programming method

Despite the popularity of the two-level programming method used in transportation or telecommunication network optimisation, it is worth noting that it is limited to the two optimisation criteria considered. Therefore, it is reasonable to propose a different approach to CNI optimisation that does not have this limitation. In another work (Guze, 2014a,b,c), the author suggested using the binary knapsack problem and solving it using the evolutionary SPEA algorithm when optimising reliability problems of complex technical systems.

Due to the discrete nature of the structure describing the CNI, it is reasonable to use this approach in this case as well. Then, multi-objective optimisation of CNI can be solved for various problems, not only those related to reliability. This approach is a universal tool and can prove very helpful in the optimal design of all CNIs, considering various criteria, such as travel time, capacity, or construction and maintenance costs.

In general, the knapsack problem is an example of a combinatorial optimisation problem that has been studied since 1897. The principle of the knapsack problem is based on a set of elements, each of which has a mass and a value. To solve this problem, the elements must be selected so that their total weight is less than or equal to a given limit, and their total value is maximised. This definition of the knapsack problem makes it a subset of NP-hard problems. This means that a non-polynomial algorithm exists to solve it. Several modifications have been used to improve the solvability of this problem, the most effective of which is the binary knapsack form (or the 0–1 knapsack problem). This modification allows the knapsack problem to be formulated as a multi-objective optimisation problem.

Essentially, the 0–1 knapsack problem, as in the general approach, is based on a set of items that have weights and profits associated with each item and an upper bound on the knapsack capacity. The main goal is to find a subset of items that maximises profits, and all selected items fit into the knapsack, i.e., the total weight does not exceed a given capacity. This one-dimensional optimisation problem can be directly extended to the multi-objective case, allowing the use of any number of knapsacks depending on the number of criteria. Formally, the 0–1 multi-objective knapsack problem can be defined as follows (Zitzler and Thiele, 1999; Guze, 2015).

For a given set of m elements and a set of n knapsacks, and $p_{i,j}$ – the profit from element j associated with knapsack i, $w_{i,j}$ – the weight of element j associated with backpack i, c_i – the capacity of backpack i, one must find the vector $\mathbf{x} = (x_1, x_2, \ldots, x_m) \in \{0,1\}^m$ such that:

$$\forall i \in \{1,2,\ldots,n\} : \sum_{j=1}^{m} w_{i,j} \cdot x_j \leq c_i \tag{4.21}$$

and for which $f(\mathbf{x}) = (f_1(\mathbf{x}), f_2(\mathbf{x}), \ldots, f_n(\mathbf{x}))$ is the maximum, where

$$f_i(\mathbf{x}) = \sum_{j=1}^{m} p_{i,j} \cdot x_j \tag{4.22}$$

and $x_j = 1$ if and only if element j is selected.

This problem formulation requires a binary representation of its structure and indicates a solution using evolutionary algorithms. These include specific methods such as genetic algorithms, genetic programming, and evolutionary strategies. Their common feature is a workflow based on the theory of natural selection and evolution. Drawing on biological analogies, we search for solutions based on an initial population, from which we select individuals (solutions) best suited (selection) to the optimisation task conditions and subject them to crossover or mutation operations. Binary notation in these algorithms is a natural notation.

To solve this optimisation problem, one of the previously mentioned evolutionary algorithms can be used. In this case, the SPEA algorithm is proposed (Guze, 2014b,c). To take advantage of the advantages offered by SPEA, the following notations should be introduced (Guze, 2014b,c; Zitzler and Thiele, 1999):

t – generation number,
P_t – population in generation t,
$\bar{P}_t$ – outer set in generation t,
$\bar{P}'$ – temporary outer set,
P' – temporary population.

Additionally, the following input parameters are required:

N – population size,
$\bar{N}$ – the largest size of the outer set,
T – the maximum number of generations,
p_c – crossover probability,
p_m – mutation probability,
A – the set of non-dominated solutions.

Then, the Strength Pareto Evolutionary Algorithm is performed in the following general steps presented in Algorithm 4.1.

Algorithm 4.1: SPEA general Steps

1. Initialisation
2. Completing the outer set with a fixed maximum size N
3. Computing the fitness function
4. Selection
5. Recombination
6. Mutation
7. Termination

The operations performed in individual steps are described in Algorithm A.7.

The proposed method is universal, enabling solutions to discrete decision-making problems regardless of the CNI model under consideration (the general one, presented in Figures 2.2 and 2.3). Considering its universality, it should be emphasised that the proposed method, together with the parallel-series network model (Figure 2.2) and formulas 2.47–2.49, allows for discrete optimisation of this network, including reliability problems. However, it is not limited to this. It can be used to optimise cost and flow between nodes for the model shown in Figure 2.3, and for the general model, as well as the entire network.

4.3.2 CONTINUOUS APPROACH TO MULTI-CRITERIA OPTIMISATION

In this section, we present optimisation methods for CNIs. Firstly, we will present bi-level programming, which is the most common multi-criteria optimisation method for transportation and telecommunications networks. However, its limitation to two criteria serves as an example to demonstrate the potential of the binary knapsack problem for multi-criteria optimisation.

As mentioned in Chapter 1, weighted graphs are those that take into account the weights of vertices or edges expressed as follows: for vertices $w_v : V(G) \to R > 0$, for edges $w_e : E(G) \to R > 0$. As can be seen, these are certain functions from the set of vertices or edges of the graph G, respectively, as a set of real values. One way to optimise these weights, depending on the problems to be solved, are known and currently used algorithms for searching, among others (Leeuwen, 1986; Cormen et al., 2009):

maximum flow – Ford–Fulkerson algorithm;
cheapest maximum flow – Klein algorithm;
shortest paths in a graph – Dijkstra and Bellman–Ford algorithms;
smallest spanning tree – Prim and Kruskal algorithms.

In the case of CNI reliability, a good description of the weights of both vertices and edges is the average failure rate of these elements, as proposed in Iida (1999).

However, the complexity of CNI problems and their often associated dynamism mean that a number description is insufficient for detailed analysis, simulation, or optimisation. Therefore, the concept of determining weights in CNIs using continuous weighting functions describing the relationships between one, two, or more parameters is justified, as defined in Chapter 2.

Back in 2015, the author showed its potential for multi-criteria approaches to technical system reliability.

For weighting functions defined in this way, it is possible to search for their extrema and thus conduct multi-objective optimisation. This can be solved using methods applicable to continuous functions. We can distinguish methods (Dixit, 1990; Findeisen et al., 1980; Ray, 2005):

1. Unconstrained optimisation:
   Direct search (sometimes also called simple search methods): Nelder–Mead simplex method (Nelder and Mead, 1965);
   Gradient optimisation: Cauchy steepest descent, Newtonian, Marquardt;
2. Constrained optimisation:
   Penalty and barrier function method;
   Rosen's method

To be able to use the above methods, the objective function should be a nonlinear real function of the vector $\mathbf{x}$, for which we are looking for the optimal point $\mathbf{x}*$ in the form:

$$f(\mathbf{x}) = f(\mathbf{x}*) + \nabla f(\mathbf{x}*)^T(\mathbf{x}-\mathbf{x}*) + \frac{1}{2}(\mathbf{x}-\mathbf{x}*)^T\nabla^2 f(\mathbf{x}*)(\mathbf{x}-\mathbf{x}*) + O_3(\Delta\mathbf{x}) \tag{4.23}$$

where $\nabla f(\mathbf{x}*)$ denotes the gradient of this function at the point $\mathbf{x}*$.

For the additional differential function of the point $\mathbf{x}*$ will be optimal, respectively, under the following conditions:

necessary – stationarity: $\nabla f(\mathbf{x}*) = 0$;
undertaking to achieve the minimum: $\nabla^2 f(\mathbf{x}*)$ is positively corporatised;
preparing for the minimum: $\nabla^2 f(\mathbf{x}*)$ is negatively corporatised.

Details of the methods mentioned above are discussed in Ray (2005). Examples of their application to engineering problems are also presented there.

The most widely used is the Nelder–Mead method, in part because it eliminates the need to compute derivatives. However, the method converges slowly, meaning that many iterations are required to find a solution close to the optimum. The gradient methods mentioned above converge much faster, but they also have the disadvantage of requiring knowledge of the derivative, a second-order derivative. Finding this derivative is often nearly impossible analytically in nonlinear problems and very time-consuming in the case of numerical solutions. Simple search methods allow for the use of a "simulate-optimise" approach.

It should be noted that the proposed in Section 2.2 weighting functions expand the possibilities of solving optimisation and simulation problems beyond the space of discrete solutions.

Firstly, we revisit fundamental concepts from the differential and integral calculus of functions of one and multiple variables (Cui et al., 2007), which underpin the main results presented. This section discusses four well-established results. The first two theorems pertain to the mean values of integrals, while the subsequent theorems address the existence of global extreme in functions of one and two variables. The first theorem is outlined as follows (Salas et al., 2007):

Theorem 4.1 *(Mean-value theorem for integrals) If f is continuous on $[a,b]$, then there is at least one number c in $[a,b]$ for which $\int_a^b f(x)dx = f(c)(b-a)$.*

This number $f(c)$ is called the average value (or mean value) of f in $[a,b]$.

The adequate theorem for the two-variable function is expressed as follows:

Theorem 4.2 *(Mean value condition) There is a point (x_0, y_0) in Ω for which $\iint\limits_{\Omega} f(x,y)dxdy = f(x_0,y_0) \cdot |\Omega|$, where $|\Omega|$ is the area of Ω. We call $f(x_0,y_0)$ the average value f on Ω.*

Suppose the form of the function f and its specificity range are known. In that case, the average value of this integral is known without calculating the integral, the same way we should think about the double integral in Theorem 4.2.

Next, the following two extreme-value theorems exist (Salas et al., 2007).

Theorem 4.3 *If f is continuous on a bounded closed interval $[a,b]$, then f takes on both a maximum value M and a minimum value m on that interval.*

Theorem 4.4 *If f is continuous on a closed and bounded set D, then f takes on an absolute maximum and an absolute minimum on that set.*

Generally, Theorems 4.1–4.4 quoted above are of great importance for the results of Section 4.3. On account of these theorems, it is possible to reduce the newly defined weights expressed in continuous, integrable, and differentiable functions to the currently known discrete weight functions. The introduced weighted functions, considering Theorems 4.1–4.4, allow us to formulate the following algorithms (see Algorithms 4.2–4.5).

The resolution of the optimisation tasks detailed in Eqs. (4)–(7) involves determining the values of decision variables that either maximise or minimise the objective function. The optimal values for the weight functions, as defined in Definitions 1 and 2, are local maxima or minima. If these are insufficient, the possibility of them representing global extremes must be ascertained according to Theorems 4.3 and 4.4. Should these values remain indeterminate, the integral mean values must be computed using Theorems 4.1 and 4.2. Consequently, the introduction of specific procedures is required to facilitate this process.

Algorithm 4.2: Minimisation-CEWG

1. Input: $G = (V, E, W_E(x_1, x_2, \ldots, x_k))$, $|E| = m$, $|V| = n$, $k \in N$
2. Output: minimal weights of edges in graph G
3. for all edges $e_i \in E$, $i = 1, 2, \ldots, m$ {
4. if ($W_{e_i}(x_1, x_2, \ldots, x_k) \in C^2$ and is an integrable function) {
5. if (exists a local minimum of $W_{e_i}(x_1, x_2, \ldots, x_k)$) {
6. return $w_{e_m}^{\min}(x_1, x_2, \ldots, x_k)$:=min$(W_{e_i}(x_1, x_2, \ldots, x_k))$;
7. } elseif (exists a global minimum of $W_{e_i}(x_1, x_2, \ldots, x_k)$) {
8. return $w_{e_m}^{\min}(x_1, x_2, \ldots, x_k)$:=min $value(W_{e_i}(x_1, x_2, \ldots, x_k))$;
9. } else return $w_{e_m}^{\min}(x_1, x_2, \ldots, x_k)$:= integral mean-value of $W_{e_i}(x_1, x_2, \ldots, x_k)$;
10. }
11. return $W_E^{\min}(x_1, x_2, \ldots, x_k) = \{w_{e_1}^{\min}(x_1, x_2, \ldots, x_k), w_{e_2}^{\min}(x_1, x_2, \ldots, x_k), \ldots, w_{e_m}^{\min}(x_1, x_2, \ldots, x_k)\}$
12. }

Algorithm 4.3: Minimisation-CVWG

1. Input: $G = (V, E, W_V(x_1, x_2, \ldots, x_k))$, $|E| = m$, $|V| = n$, $k \in N$
2. Output: minimal weights of vertices in graph G
3. for all edges $v_i \in V$, $i = 1, 2, \ldots, n$ {
4. if ($W_{v_i}(x_1, x_2, \ldots, x_k) \in C^2$ and is an integrable function) {
5. if (exists a local minimum of $W_{v_i}(x_1, x_2, \ldots, x_k)$) {
6. return $w_{v_n}^{\min}(x_1, x_2, \ldots, x_k)$:=min$(W_{v_i}(x_1, x_2, \ldots, x_k))$;
7. } elseif (exists a global minimum of $W_{v_i}(x_1, x_2, \ldots, x_k)$){
8. return $w_{v_n}^{\min}(x_1, x_2, \ldots, x_k)$:=min $value(W_{v_i}(x_1, x_2, \ldots, x_k))$;
9. } else return $w_{v_n}^{\min}(x_1, x_2, \ldots, x_k)$:= integral mean-value of $W_{v_i}(x_1, x_2, \ldots, x_k)$;
10. }
11. return $W_V^{\min}(x_1, x_2, \ldots, x_k) = \{w_{v_1}^{\min}(x_1, x_2, \ldots, x_k), w_{v_2}^{\min}(x_1, x_2, \ldots, x_k), \ldots, w_{v_m}^{\min}(x_1, x_2, \ldots, x_k)\}$
12. }

Algorithm 4.4: Maximisation-CEWG

Input: $G = (V, E, W_E(x_1, x_2, \dots, x_k))$, $|E| = m$, $|V| = n$, $k \in N$
Output: maximal weights of edges in graph G
for all edges $e_i \in E$, $i = 1, 2, \dots, m$ {
if $(W_{e_i}(x_1, x_2, \dots, x_k) \in C^2$ and is an integrable function) {
if (exists a local maximum of $W_{e_i}(x_1, x_2, \dots, x_k)$) {
return $w_{e_m}^{\max}(x_1, x_2, \dots, x_k) := \max(W_{e_i}(x_1, x_2, \dots, x_k))$;
} elseif (exists a global maximum of $W_{e_i}(x_1, x_2, \dots, x_k)$){
return $w_{e_m}^{\max}(x_1, x_2, \dots, x_k) := \max value(W_{e_i}(x_1, x_2, \dots, x_k))$;
} else return $w_{e_m}^{\max}(x_1, x_2, \dots, x_k) :=$ integral mean-value of $W_{e_i}(x_1, x_2, \dots, x_k)$;
}
return $W_E^{\max}(x_1, x_2, \dots, x_k) = \{w_{e_1}^{\max}(x_1, x_2, \dots, x_k), w_{e_2}^{\max}(x_1, x_2, \dots, x_k), \dots, w_{e_m}^{\max}(x_1, x_2, \dots, x_k)\}$
}

Algorithm 4.5: Minimisation-CVWG

Input: $G = (V, E, W_V(x_1, x_2, \dots, x_k))$, $|E| = m$, $|V| = n$, $k \in N$
Output: maximal weights of vertices in graph G
for all edges $v_i \in V$, $i = 1, 2, \dots, n$ {
if $(W_{v_i}(x_1, x_2, \dots, x_k) \in C^2$ and is an integrable function) {
if (exists a local maximum of $W_{v_i}(x_1, x_2, \dots, x_k)$) {
return $w_{v_n}^{\max}(x_1, x_2, \dots, x_k) := \max(W_{v_i}(x_1, x_2, \dots, x_k))$;
} elseif (exists a global maximum of $W_{v_i}(x_1, x_2, \dots, x_k)$){
return $w_{v_n}^{\max}(x_1, x_2, \dots, x_k) := \max value(W_{v_i}(x_1, x_2, \dots, x_k))$;
} else return $w_{v_n}^{\max}(x_1, x_2, \dots, x_k) :=$ integral mean-value of $W_{v_i}(x_1, x_2, \dots, x_k)$;
}
return $W_V^{\max}(x_1, x_2, \dots, x_k) = \{w_{v_1}^{\max}(x_1, x_2, \dots, x_k), w_{v_2}^{\max}(x_1, x_2, \dots, x_k), \dots, w_{v_m}^{\max}(x_1, x_2, \dots, x_k)\}$
}

4.4 EXAMPLES OF APPLICATION

In this section, we present two possibilities of applying the theories introduced in Section 3.2

4.4.1 MULTI-CRITERIA OPTIMISATION OF COMPLEX TECHNICAL NETWORK

An example of the application of the solution proposed and described in Section 4.3.2, consisting in reducing the problem to a binary knapsack problem, was presented by the author in Guze (2015). This solution was based on a modification of the solution offered by the PISA project platform, which is developed at the Zurich University of Technology (Zitzler and Thiele, 1999).

Based on Guze (2014a), the possibility of applying the proposed method to the reliability analysis of parallel-series technical systems is presented below. For this approach, we make the following assumptions:

- n – number of system's components;
- E_{ij}, $i = 1,2,\ldots,k_m$, $j = 1,2,\ldots,l_i$ – highlighted elements of systems;
- T_{ij} – independent random variables representing the lifetime of components E_{ij}, $i = 1,2,\ldots,k_m$, $j = 1,2,\ldots,l_i$;
- $R_{ij}(t) = P(T_{ij} > t)$, $t \in < 0,+\infty)$ – reliability function of system's elements E_{ij}, $i = 1,2,\ldots,k_m$, $j = 1,2,\ldots,l_i$;
- $F_{ij}(t) = 1 - R_{ij}(t) = P(T_{ij} \leq t)$, $t \in < 0,+\infty)$ – distribution of the lifetimes T_{ij} system components E_{ij}, $i = 1,2,\ldots,k_n$, $j = 1,2,\ldots,l_i$.

Furthermore, we assume that the elements E_{ij}, $i = 1,2,\ldots,k_n$, $j = 1,2,\ldots,l_i$, form parallel subsystems S_i, $i = 1,2,\ldots,k_n$ and these in turn form a series system.

We assume that the system under consideration (see Figure 4.2) is a two-state parallel-series structure that has three main subsystems S_i, $i = 1,2,3$. Each subsystem is parallel and consists of as many elements as possible, which are selected to

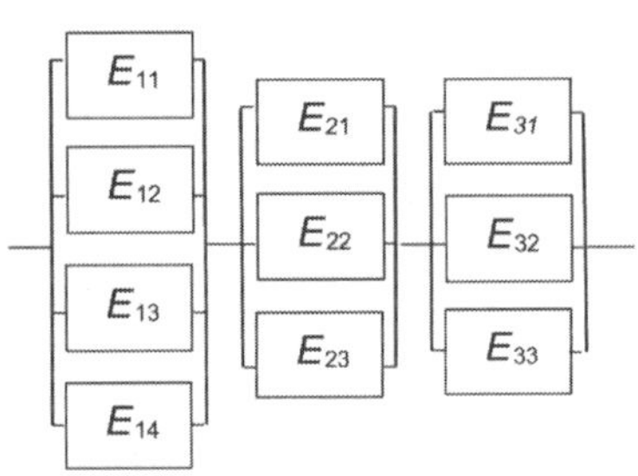

Figure 4.2 Structural model of complex technical system

Table 4.1
Initial Poplulation for SPEA

ID	Chromosom	MTTF [h]	Cost [USD]	ID	Chromosom	MTTF [h]	Cost [USD]
0	0111000010	5460	42100	25	1011110001	4620	41500
1	0011110111	4040	35000	26	0010111010	5260	40400
2	0111111101	5010	37900	27	1110011110	5780	40400
3	0101101011	4490	31100	28	0011110111	4040	35000
4	1011000001	4760	39600	29	1111001101	5590	37600
5	1011010111	4010	35900	30	1111110100	4960	42400
6	1011000000	5630	46100	31	0101110100	4960	42400
7	0101101010	4830	32000	32	0100001000	7470	52200
8	1001000011	5260	37800	33	0010110010	4840	44900
9	0101000111	4840	32900	34	1100011010	5780	40400
10	0110000111	5600	43700	35	0000111000	5840	46900
11	0001110110	4910	41500	36	1011101010	4280	32900
12	1010011000	4870	39700	37	1000010001	5740	47100
13	1111011001	4480	32300	38	1011101000	4860	39400
14	1001001110	5470	35100	39	1111011111	4430	31400
15	1001111000	4830	38800	40	0100110110	5390	44000
16	1001111111	4430	31400	41	0101111110	5300	37900
17	0010010011	5020	44000	42	0110101100	5340	41900
18	0110010001	5070	44900	43	1110010000	5000	43300
19	1101110111	4040	35000	44	0000111000	5840	46900
20	1010110101	4550	44900	45	0000011101	5190	38600
21	0100101010	5840	40100	46	1001111011	4430	31400
22	0110011100	5280	42200	47	0000000100	7830	59600
23	1000100110	6090	46800	48	0110010010	5360	44900
24	0011010001	4590	42400	49	1100001100	6530	44100

Source: Our work based on Guze (2015)

improve the system's resilience. The maximum numbers of elements in individual subsystems are as follows:

four – for S_1 subsystem
three – for every subsystem S_2 and S_3

As previously assumed, each system element is bi-state, meaning it is either operational or faulty. The main characteristics of these elements are their MTTF and installation costs, examples of which are presented in Table 4.1.

In real-world applications, the discrete optimisation method proposed in Section 4.3 can be used to answer the question: *How should a new system be created or an existing one redesigned to maximise the time to failure while minimising installation and operating costs?*

This means that the goal of solving the problem is to maximise the system's failure time and minimise its installation and operating costs. This is a classic two-objective optimisation. This problem can be solved by reducing it to the problem of binary knapsacks problem.

This is a classical two-objective optimisation. Based on the results presented in Section 4.3, this problem can be solved by reducing it to the problem of packing two binary knapsacks. In this solution, we make the following assumptions:

c_i – time to failure of the designed system
$p_{i,j}$ – profit equal to the lifetime of individual components
$w_{i,j}$ – the cost of installation and use of an component of a given type

Furthermore, suppose that the chromosome shown in Figure 4.3 represents an example of the structure of the entire system. A gene value of one indicates that the element is included in the knapsack. A gene in this chromosome equals zero if the component designated by it is not in the knapsack. The length of the chromosome is the number of elements being examined (the number of system elements).

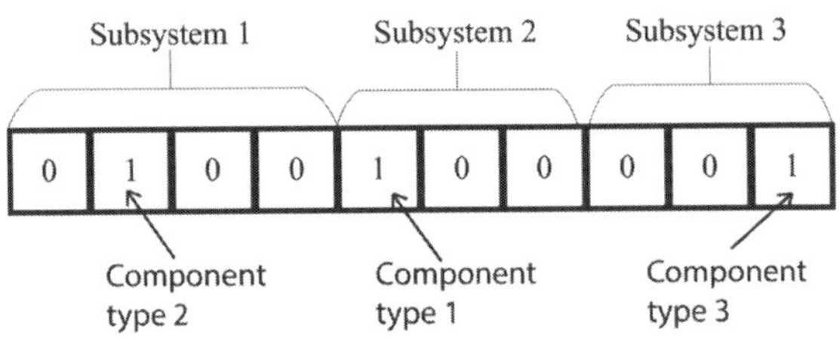

Figure 4.3 An example chromosome representing a technical system

For the system shown in Figure 4.2, and considering the data from Table 4.1, the time to failure is optimised for the lowest cost. A solution to the optimisation problem of a complex technical system was proposed, taking into account the two criteria mentioned above.

The solution to improve the existing system was implemented using the theory presented in Section 4.3, using a computer program written by the author based on the source code developed as part of the PISA project (details in Guze (2014a, 2015). To solve this optimisation problem, five cases were assumed, each with different program startup parameters:

Case 1: mutation˙type 1, recombination˙type 1.
Case 2: mutation˙type 1, recombination˙type 2.
Case 3: mutation˙type 2, recombination˙type 2.
Case 4: mutation˙type 2, recombination˙type 1.
Case 5: mutation˙type 2, recombination˙type 2, mu 20, lambda 20.

The initial population in all of the above cases is the same. Details are provided in Table 4.1. As can be seen, the types of mutation and chromosomal recombination differ in cases 1–4. Furthermore, in Case 5, these parameters are the same as in case 3, but the program startup parameters, "mu" and "lambda", differ.

The program results for Cases 1–5 are shown in Tables B.1–B.5 in Appendix B. Cells containing chromosomes forming the Pareto front are written down the tables.

With reference to the results in Tables B.1–B.5, the selected Pareto-optimal structures for Cases 1–5 of the considered system are presented in Figure 4.4.

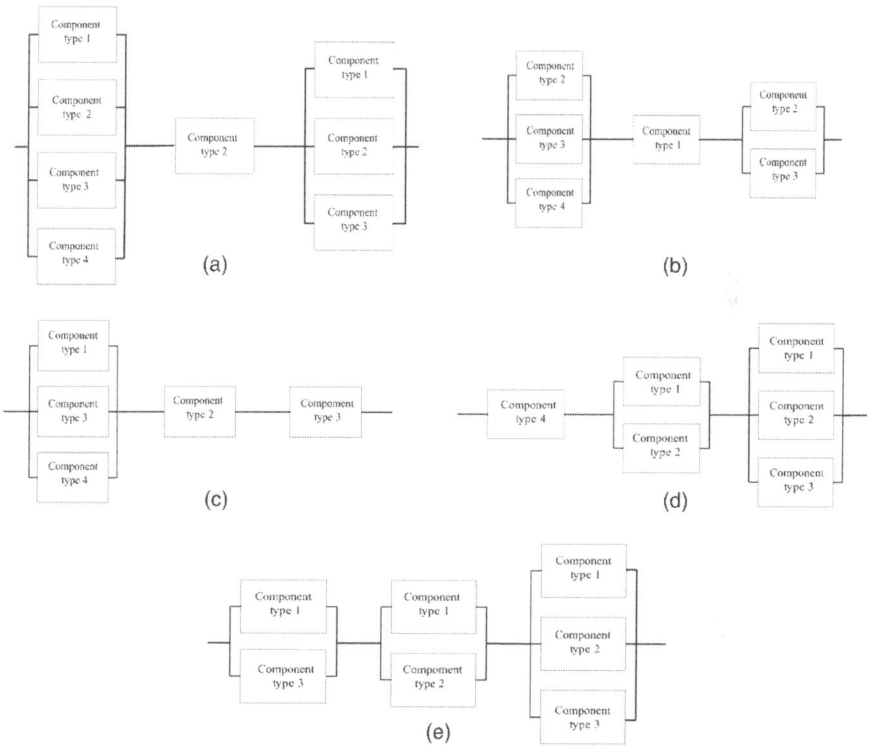

Figure 4.4 (a) - system with eight subsystems; (b) and (d) - system with six subsystems in different configurations; (c) - system with five subsystems; (e) - system with seven subsystems considered cases

4.4.2 MULTI-CRITERIA OPTIMISATION OF TRANSPORTATION NETWORK

This section illustrates the theory presented in Sections 2.2 and 4.2 through an academic example, demonstrating the capabilities of the defined graph families for modelling and optimising transportation networks. Consider an exemplary graph structure of a transportation network as shown in Figure 4.2. This network includes a set of vertices $V = \{A,B,C,D,E,F,G\}$ and a set of edges $E = \{e1 = (A,B), e2 = (B,F), e3 = (A,C), e4 = (A,E), e5 = (C,D), e6 = (C,F), e7 = (B,C), e8 = (D,F), e9 = (D,E), e10 = (E,F), e11 = (E,G)\}$.

In accordance with Definitions 2.1, we define the continuous weighted functions for each edge $R_{e_i}(t,c)$, where $e_i \in E$ and $i = 1,\ldots,11$. These functions characterise the reliability of the network's arcs, which varies over time t and influences the generation of traffic flow density c. This setup allows us to explore dynamic changes in the network and provides a framework for optimising network performance based on varying conditions.

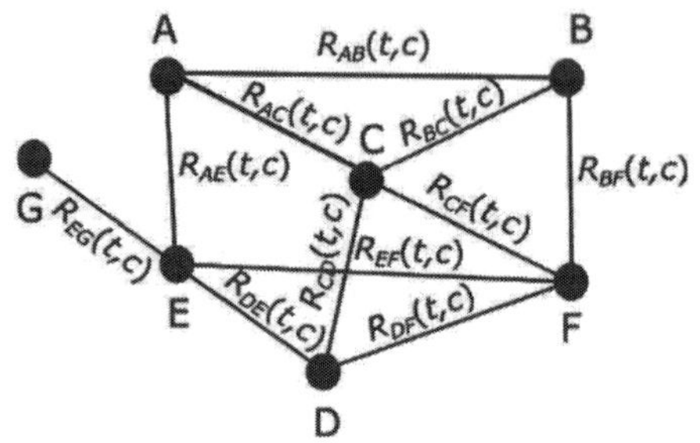

Figure 4.5 Part of the urban transportation network with seven nodes

Figure 4.7 illustrates a segment of an urban transportation network, focusing on reliability as it varies with time and traffic flow density. To model this scenario effectively, we employ the multivariate Weibull distribution . The general formula for the multivariate Weibull distribution (Crowder, 1989; Lee, 1979; Lu and Bhattacharyya, 1990; Roy and Mukherjee, 1989) is expressed as:

$$R(\mathbf{x}) = \exp\{-\left[\left(\frac{x_1}{\lambda_1}\right)^{\frac{\gamma_1}{\alpha}} + \left(\frac{x_2}{\lambda_2}\right)^{\frac{\gamma_2}{\alpha}} + \ldots + \left(\frac{x_n}{\lambda_n}\right)^{\frac{\gamma_n}{\alpha}}\right]^{\alpha}\} \tag{4.24}$$

where $\alpha \in (0,1>$, $\mathbf{x} = (x_1, x_2, \ldots, x_n)$, $\gamma_i \in (0,\infty)$, $\lambda_i \in (0,\infty)$ for all $i = \{1,2,\ldots,n\}$.

In the case considered, we assume that $x1 = t$ and $x2 = c$ in Eq. 4.24. Thus, we obtain the following formula for the edge reliability functions:

$$R_{e_i}(t,c) = \exp\left\{-\left[\left(\frac{t}{\lambda_1}\right)^{\frac{\gamma 1}{\alpha}} + \left(\frac{c}{\lambda_2}\right)^{\frac{\gamma_2}{\alpha}}\right]^{\alpha}\right\} \tag{4.25}$$

where $R_{e_i}(t,c)$ is the reliability continuous weighted functions for every edge $e_i \in E$, and $i = 1,\ldots,11$, $\alpha \in (0,1]$, $y1$, $y2 \in (0,\infty)$, $\lambda 1, \lambda 2 \in (0,\infty)$, t, $c \in < 0,\infty)$.

In Table 4.2, we present the assumed values for the parameters α, $y1$, $y2$, $\lambda 1$ and $\lambda 2$, required for Eq 4.25.

Substituting the values from Table 4.2 into Eq. 4.25 for individual network arcs, we obtain the following forms of reliability functions:

$$R_{e_1}(t,c) = \exp\{(-(t^2 + c^{20})^{10})\} \tag{4.26}$$

$$R_{e_2}(t,c) = \exp\{(-(t^2 + c^{20})^{10})\} \tag{4.27}$$

Table 4.2
Values of the Multivariate Weibull Distribution

edge	α	y1	y2	$\lambda 1$	$\lambda 2$
$e1$	10	5	0.5	1	1
$e2$	10	5	0.5	1	1
$e3$	10	5	0.5	1	1
$e4$	10	5	0.5	1	1
$e5$	10	5	0.5	1	1
$e6$	8	2	0.5	1	1
$e7$	10	5	0.5	1	1
$e8$	10	5	0.5	1	1
$e9$	10	5	0.5	1	1
$e10$	8	2	0.5	1	1
$e11$	5	5	0.5	1	1

$$R_{e_3}(t,c) = \exp(-\left(t^2 + c^{20}\right)^{10}) \tag{4.28}$$

$$R_{e_4}(t,c) = \exp(-\left(t^2 + c^{20}\right)^{10}) \tag{4.29}$$

$$R_{e_5}(t,c) = \exp(-\left(t^2 + c^{20}\right)^{10}) \tag{4.30}$$

$$R_{e_6}(t,c) = \exp(-\left(t^4 + c^{16}\right)^{8}) \tag{4.31}$$

$$R_{e_7}(t,c) = \exp(-\left(t^2 + c^{20}\right)^{10}) \tag{4.32}$$

$$R_{e_8}(t,c) = \exp\left(-\left(t^2 + c^{20}\right)^{10}\right) \tag{4.33}$$

$$R_{e_9}(t,c) = \exp(-\left(t^2 + c^{20}\right)^{10}) \tag{4.34}$$

$$R_{e_{10}}(t,c) = \exp(-\left(t^4 + c^{16}\right)^{8}) \tag{4.35}$$

$$R_{e_{11}}(t,c) = exp\left(-\left(t^{10} + c\right)^{5}\right) \tag{4.36}$$

Reliability structures for transportation networks represented by connected simple graphs are well documented in the literature (Guze, 2019b). Given the specific assumptions for the analysed segment of the network, it must be categorised as a serial transportation network for subsequent calculations.

We can assume that this exemplary transportation network is classified as serial in the context of edges. Furthermore, the reliability function for this type of transport network is defined as per the following equation:

$$R(\mathbf{x}) = \prod_{i=1}^{|E|} R_{e_i}(\mathbf{x}) \tag{4.37}$$

where $\mathbf{x} = (x_1, x_2, \ldots, x_n)$, $R(\mathbf{x})$ is the transportation network reliability function, and $R_{e_i}(\mathbf{x})$ is the edge reliability function.

Based on earlier definition and considering the formulae of the reliability functions in Eqs. 4.26–4.37, we can define the transportation network's series two-variable reliability function as follows:

$$R(t,c) = \exp\left(-2 * \left(t^4 + c^{16}\right)^8\right) * \exp\left(-8 * \left(t^2 + c^{20}\right)^{10}\right) * \exp\left(-\left(c^{10} + t\right)^5\right) \tag{4.38}$$

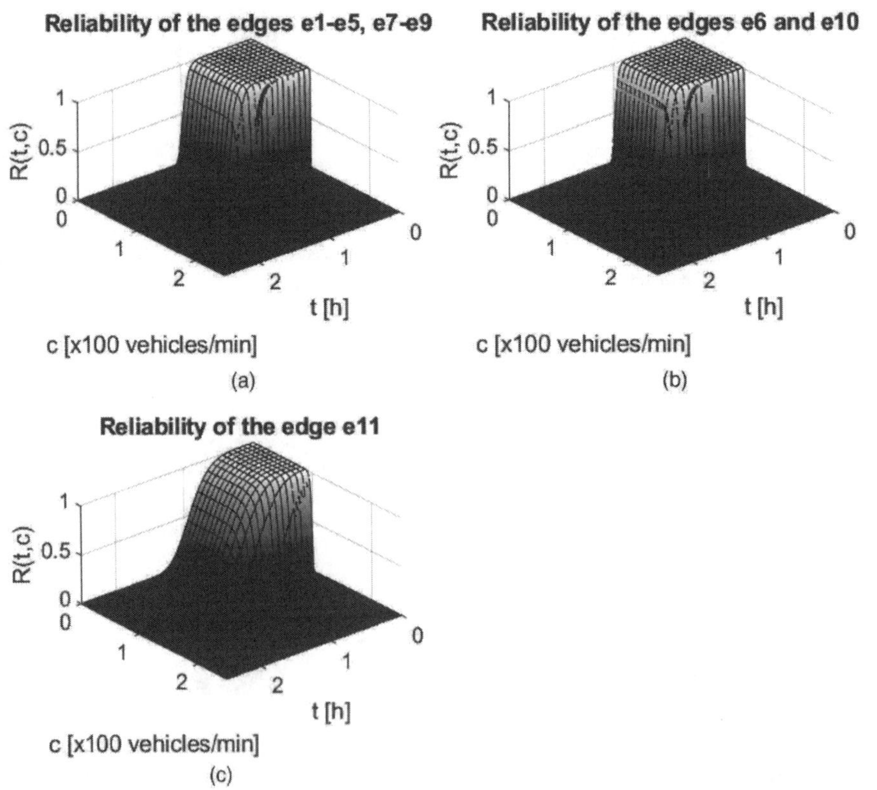

Figure 4.6 (a) - graph of the reliability function of the edges e1-e5 and e7 - e9; (b) - graph of the reliability function of the edges e6 and e10; (c) - graph of the reliability function of the edge e11

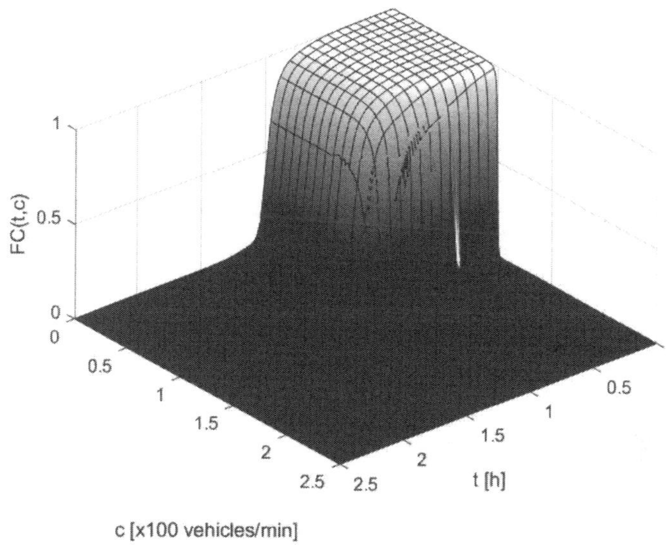

Figure 4.7 Goal function for transportation function reliability optimisation

The reliability function, using assumed academic values, is presented in Figure 4.6.

Using the MATLAB software R2023a, we obtain the optimal solutions for the function in Eq. 4.38: $t = 1.0500$, $c = 0.7700$, and $R(t,c) = 9.3884e-16$. Next, we consider formula 4.38 and reliability functions 4.26–4.34 and express the goal function as follows:

$$F_C(t,c) = \frac{1}{5} \cdot \exp\left(-\left(t^4 + c^{16}\right)^8\right) + \left(\frac{3}{5} \cdot \exp\left(-\left(t^2 + c^{20}\right)^{10}\right)\right) + \frac{1}{5} \cdot \exp\left(-\left(c^{10} + t\right)^5\right) \tag{4.39}$$

The function is illustrated in Figure 4.7. Similarly, we utilised MATLAB software R2023a to derive the optimal solutions for the function specified in Eq. 4.39. The resulting solution values are $t = 0.3218$ and $c = 10.7434$, with $FC(t,c) = 0.0$.

The optimisation was conducted with the constraint that the decision variables t and c must be greater than zero. In the example, we were interested in finding the minimum reliability function of the transportation network – that is, we minimised its value. We used this to show at what time and traffic intensity the affected part of the transportation network will be congested. The result is somewhat surprising because it indicates that congestion will occur after $0.77[h]$ and at a traffic density of over $10,000$ cars per minute. Of course, these results are based on assumptions made about the parameters of the Weibull distribution.

This example is merely one of many potential applications of the new graph families defined in Section 2.2. It demonstrates the utility of these concepts but does not fully explore the extensive possibilities offered by the theory proposed in Sections 2.2 and 3.2.

It should be emphasised that without defining new graph families, the reliability analysis shown in the example above could not be performed due to the shortcomings of such models. This is demonstrated in the literature review on available graph models for describing transportation networks. The methodological shortcomings identified there limited the possibilities of modelling time-varying phenomena, including driving speed, capacity, number of accidents, and costs. They also did not provide the opportunity to study the relationships among phenomena affecting traffic flow.

The introduction of new graph families provides these possibilities. Moreover, looking at the sample results of this section, it is possible to propose routing optimisation in electronic maps in the future due to travel safety related to time and cost. Although the example is academic, it shows the possibilities of the theory proposed in the book.

5 Summary and Final Conclusions

This book defines critical network infrastructures (CNIs), which form the basis for the operation of all other critical infrastructures. This is due to technological developments, which have resulted in the transfer of all public, financial, and educational services to the digital world. Simultaneously, it increases pressure on the reliable operation of energy, ICT, and transportation systems.

Therefore, this book presents the application of selected methods, algorithms, and characteristics of graph theory and game theory to the modelling, analysis, and optimisation of critical network infrastructures. The author places particular emphasis on solving problems related to their reliability, vulnerability, and resilience.

Therefore, the book provides a comprehensive description of the reliability of complex, CNIs, utilising concepts from graph and game theories. It is done from a structural and functional perspective and in terms of components and whole system lifetime. Firstly, for this purpose, a total of eight network reliability structures are defined. Based on these results, two methods for describing the reliability function of a CNI are proposed, using both the structural function and the lifetime function. This part of the monograph assumes that CNIs, from the perspective of their functioning, are complex technical systems.

Additionally, new measures for assessing the vulnerability of the CNI are defined, using related concepts to describe dominating sets in graphs. Research results in this area include the vulnerability and resilience of the CNI to the unreliability of individual edges in terms of traffic control, both in the simple graph and vertex-weighted graph versions. The second type of vulnerability considered in this book refers to the sensitivity of the network under consideration to increasing the number of vertices.

Another significant contribution of this work is the development of two methods for optimising CNIs: a discrete variant, which is natural for graph theory, and a continuous variant, which requires defining new classes of weighting functions in the graph as functions of multiple variables.

The first approach proposed the reducibility of discrete optimisation problems for CNIs to the binary knapsack problem, which in turn requires the use of genetic algorithms. This also determined the need to propose a chromosomal encoding of the network.

The second approach allows the use of linear and nonlinear methods for the analysis and optimisation of critical network infrastructures, which was previously impossible. The correctness of the introduced description is confirmed by four algorithms designed explicitly for searching minima and maxima. These provide

DOI: 10.1201/9781003661450-5

solutions if the functions chosen to describe the behaviour of individual CNI elements meet the assumptions, but do not have the appropriate optimal values.

The proposed methods and algorithms based on game theory are used to build functional and decision-making models. They also enable the simulation analysis of CNIs, particularly those focused on the resilience of individual elements within these networks.

The practical application of the methods developed and presented in this monograph in the modelling, analysis, and optimisation of CNIs can be divided into:

- in the design or planning analysis of network structures;
- in the online analysis of a functioning network, for the management of data, energy, or transportation flows.

During the design phase of CNIs, the methods presented in this work enable analysis and simulations to identify and eliminate weak links and bottlenecks. It allows the development of a network structure that is resilient to various undesirable factors. This approach to designing CNIs minimises the costs associated with congestion or shortages, as well as the potential for subsequent network upgrades.

In the case of online analysis, the tools introduced in this work enable the even distribution of data packets, traffic flow, or electricity – depending on the selected network – in the event of a connection or node failure, by identifying nodes within the network that are crucial for ensuring reliable operation. This information enables network administrators and traffic engineers, particularly in cases involving intelligent power systems, traffic management, or data packet transmission, to quickly respond to the occurrence of a malfunction in a specific section of the network structure.

Relating the results obtained in the book to the current state of knowledge allows us to draw the following conclusions:

1. The novelty and originality of the tools proposed in the book, which contribute to the development of methods for modelling, analysing, and optimising critical network infrastructures, lie in:
    a. defining network infrastructure models, taking into account their various levels of representation in relation to real networks;
    b. developing new network parameters and indicators based on concepts derived from domination in graphs;
    c. defining the reliability structures of CNIs in the vertex and edge sense;
    d. developing methods for modelling the reliability of complex two-state, multi-state, and mixed technical systems for complex CNIs;
    e. defining parameters enabling the assessment of the vulnerability of CNIs in a topological and weighted approach;
    f. defining the weighting function for vertices and edges in a graph representing network infrastructure as a function of multiple variables;
    g. developing a multi-objective optimisation method for critical systems and networks, based on the reduction of this problem to the binary knapsack problem using an evolutionary algorithm and appropriate binary encoding of chromosomes for CNIs;

   h. proposing an optimisation method for weights described by multi-variable functions.
2. The proposed methods can be beneficial in online analysis of critical network infrastructures, or even earlier, at the stage of designing their topology.
3. Proposing multi-variable functions as weighting functions expands the availability of optimisation methods for network structures described by weighted graphs. It is no longer necessary to solve these types of problems using two-level programming or simpler linear programming. The existing multi-objective optimisation methodology, in its discrete variant, which utilises the binary knapsack problem, has also gained a new tool for finding solutions to well-known problems, such as the shortest path problem and the maximum flow problem.
4. This book provides conclusions that enrich the existing methodological framework for modelling and analysing CNIs. The methodology proposed here can be successfully used in both the design and operational phases of critical networks.

Despite the comprehensiveness of the presented results, several research areas remain necessary:

1. Proposing more efficient algorithms for determining dominating sets;
2. Developing a computer-aided system for the analysis, design, and optimisation of arbitrary network structures described by graphs.

A Selected Algorithms

A.1 ALGORITHMS FOR SECOND CHAPTER

Algorithm A.1: Depth First Search Algorithm

1. function DFS(adj[][], disc[], low[], visited[], parent[], AP[], vertex, V)
2. visited[vertex] = true
3. disc[vertex] = low[vertex] = time+1
4. child = 0
5. for i = 0 to V
    a. if adj[vertex][i] == true
    b. if visited[i] == false
        i. child = child + 1
        ii. parent[i] = vertex
        iii. DFS(adj, disc, low, visited, parent, AP, i, n, time+1)
        iv. low[vertex] = minimum(low[vertex], low[i])
        v. if parent[vertex] == nil and child ¿ 1
            A. AP[vertex] = true
        vi. if parent[vertex] != nil and low[i] ¿= disc[vertex]
            A. AP[vertex] = true
    c. else if parent[vertex] != i
        i. low[vertex] = minimum(low[vertex], disc[i])

Sources: https://www.hackerearth.com/

Algorithm A.2: D2.2

1. function find-bridges($adj[][], V, Edge[], E, isBridge[]$)
2. $for i = 0 \ to \ E$
    a. $adj[Edge[i].u][Edge[i].v] = adj[Edge[i].v][Edge[i].u] = 0$
3. for $j = 0$ to V
4. $visited[j] = false$
5. $Queue.Insert(Edge[i].u)$
6. $visited[Edge[i].u] = true$
7. check = false
8. $whileQueue.isEmpty() == false$
9. $x = Queue.top()$
10. $ifx == Edge[i].v$
11. $check = true$
12. BREAK
13. Queue.Delete()
14. for $j = 0$ to V
15. if $adj[x][j] == trueANDvisited[j] == false$
16. Queue.insert(j)
17. $visited[j] = true$
18. $adj[Edge[i].u][Edge[i].v] = adj[Edge[i].v][Edge[i].u] = 1$
19. if $check == false$
20. $isBridge[i] = true$

Source: https://www.hackerearth.com/practice/algorithms/graph

Algorithm A.3: Minimal Cutsets Enumeration Algorithm

1. RecursiveMerge(G, SS, n)
2. a. if (n == t) return;
    b. G = G*n; SS = SS + n;
    c. Recursively absorb (merge) redundant nodes of G into SS;
    d. if (SS is found in the hash table) return;
    e. else add SS to the hash table;
3. Output a cutset of SS;
4. for each node ni adjacent to SS
5. a. RecursiveMerge(G, SS, ni);

Algorithm A.4: D3.3A

1. For all the smallest cuts found using algorithm A.3 do
    a. If the edges are adjacent then return true;
    b. else false;
2. end for.

Algorithm A.5: D2.4

```
function find-articulation-points(adj[][], V)
count = 0
for i = 0 to V
visited[i] = false
initial-val = 0
for i = 0 to V
if visited[i] == false
DFS(adj, V, visited, i)
initial-val = initial-val+1
for i=0 to V
for j = 0 to V
visited[j] = false
copy[j] = adj[i][j]
adj[i][j]=adj[j][i]=0
nval = 0
for j= 0 to V
if visited[j] == false AND j != i
nval = nval + 1
DFS(adj, n, visited, j)
if nval ¿ initial-val
count = count + 1
for j= 0 to V
adj[i][j] = adj[j][i] = copy[j]
return count
time = 0
Source [https://www.hackerearth.com/practice/algorithms/graph]
```

Algorithm A.6: Vertex Disjoint Paths Algorithm

Input: Graph G(V,E), value of a number $\kappa(u, v)$ Output: Set of separating vertices

1. Find the minimum distances of each vertex from the target vertex using a modified breadth-first search (BFS) algorithm.
2. Select one vertex from the source vertex's neighbours that is closest to the target vertex (using the data obtained in Step 1).
3. If the distance of the neighbouring vertices of the nearest vertex to the target vertex is greater than 1, select the closest vertex adjacent to the target vertex.
4. If the path is complete, mark the used vertices so that they cannot be reused.
5. Continue until a new path is found.

Source: [https://www.hackerearth.com/]

A.2 ALGORITHM FOR FOURTH CHAPTER

Algorithm A.7: Strength Pareto Evolutionary Algorithm (SPEA)

1. Initialisation
   a. Select individual i.
   b. Eliminate the set P_0.
   c. Generate an empty outer set $\bar{P}_0$ for t=0.
2. Completing the distribution set
   a. Let $\bar{P}'=\bar{P}_t$.
   b. Copy the non-dominated elements from the equipment P_t to the operation $\bar{P}'$.
   c. Create a dominated element from the set $\bar{P}'$.
   d. Reduce the size of the set $\bar{P}'$ to N using the clustering operation, the result of which is stored in $\bar{P}_{t+1}$.

   The search for the value of the fitness function F_{fit} for elements from the sets P_t and $\bar{P}_t$ can be performed according to the procedure [Guze 2015, Zitzler et al., 1999]:

   a. The real value, the so-called strength, $S \in [0,1)$ is defined for each element $i \in \bar{P}_t$. At the same time, it is proportional to the number of elements $j \in P_t$, which represent solutions dominated by element i.
   b. The fitness of individual j is calculated as the sum of all elements from the outer set, which represents the solution dominated by element j, increased by 1. Increasing this value by 1 ensures that element $i \in \bar{P}_t$ can have a better fitness function value than an element from the set P_t, i.e.:

   $$S(i) = \frac{n}{N+1}$$

   where
   - $S(i)$ - the strength of element i,
   - n - the number of elements in the population dominated by element i.

   c. It is assumed that the fitness function of element i is equal to its strength/cardinality, i.e.

   $$F(i) = S(i)$$

3. Selection
   a. Let $P' = \emptyset$.
   b. For $i = 1,2,\ldots,k$ do
      i. Randomly select two individuals $i,j \in P_t \cup \bar{P}_t$.
      ii. If $F(i) < F(j)$ then $P' = P' \cup \{i\}$ else $P' = P' \cup \{j\}$, assuming that the fitness value is minimised.
4. Recombination
   a. Let $P'' = \emptyset$.
   b. For $i = 1,2,\ldots,\frac{N}{2}$ do:
      i. Select two individuals $i,j \in P'$ and remove them from $\bar{P}'$.
      ii. Create elements k,l by crossing individuals i,j.
      iii. Add individuals k,l to the set P'' with probability p_c, otherwise add individuals i,j to the set P''.
6. Mutation
   a. Let $P''' = \emptyset$.
   b. For each individual $i \in P''$, perform the following:
      i. Create individual j by mutating individual i with probability p_m.
      ii. Add individual j to the set P'''.
7. Finalise
   a. Let $P_{t+1} = P'''$ and $t = t+1$.
   b. If $t \geq T$, then return A – non-dominated solutions from the population P_t and terminate, otherwise return to step 2.

B Resulting Chromosomes for Example in Chapter Four

Table B.1
Results for Case 1

ID	Chromosom	MTTF [h]	Cost [USD]	ID	Chromosom	MTTF [h]	Cost [USD]
1	1011010001	5010	32700	52	1011000011	4480	33900
3	0101101011	4170	35900	54	1001111011	4230	35600
4	1100111011	4270	33100	55	1011100001	4950	33000
6	1111010111	5180	36700	56	1011100101	4950	33000
10	1001111111	4230	35600	57	0011101011	4170	35900
15	1011010001	5010	32700	59	1111110011	4620	32000
16	1001111111	4230	35600	60	0101111111	4230	35600
17	1011100001	4950	33000	61	1011000110	4440	34800
25	1111010011	5180	36700	62	1011010001	5010	32700
27	1001111111	4230	35600	64	1011010001	5010	32700
30	1011010001	5010	32700	69	1001111011	4230	35600
31	1011100001	4950	33000	72	1011000110	4440	34800
32	1001111111	4230	35600	76	1001101010	4240	37500
33	1011000111	4480	33900	79	1011010111	4650	31100
35	1001101010	4240	37500	82	1011000111	4480	33900
36	1011101010	4380	34100	83	1011000110	4440	34800
38	1011000111	4480	33900	84	1001110011	4620	32000
39	1111011111	4230	35600	88	1111011011	4230	35600
40	1111010111	5180	36700	89	1111010111	5180	36700
41	0101101011	4170	35900	92	1011000011	4480	33900
42	1111111111	4230	35600	94	1101111011	4230	35600
43	1111010000	4670	31800	95	1001101011	4170	35900
44	1011000110	4440	34800	97	1001111011	4230	35600
46	1001111011	4230	35600	98	1011000111	4480	33900
47	1001010111	4510	34500	99	1011010011	4650	31100

Source: Own work based on Guze (2015)
Note: Chromosomes constitute Pareto front ID:1, 6, 15, 17, 25, 30, 31, 40, 43, 55, 56, 62, 64, 79, 89, 99

DOI: 10.1201/9781003661450-B

Table B.2
Results for Case 2

ID	Chromosom	MTTF [h]	Cost [USD]	ID	Chromosom	MTTF [h]	Cost [USD]
2	1111100011	5120	37000	57	0111100011	5120	37000
3	0101101011	4170	35900	60	1001010011	4510	34500
4	1001010011	4510	34500	62	1001010111	4510	34500
11	0111100011	5120	37000	63	1111100011	5120	37000
13	1001011110	4300	37200	64	0111100011	5120	37000
14	1011100010	4660	33000	66	1011100110	4660	33000
16	1001111111	4230	35600	68	1111011111	4230	35600
18	1011100110	4660	33000	70	1011100110	4660	33000
23	0111100011	5120	37000	71	1001011111	4230	35600
25	1011100110	4660	33000	74	1001100111	4450	34800
28	1011010001	5010	32700	75	1011010010	4720	32700
29	1001010011	4510	34500	79	0111100011	5120	37000
31	1001011110	4300	37200	80	1001100111	4450	34800
33	1001100111	4450	34800	81	1001101111	4170	35900
34	1001010011	4510	34500	83	1011010101	5010	32700
39	1111011111	4230	35600	84	1001010011	4510	34500
40	1011100110	4660	33000	85	1011100010	4660	33000
43	1011000110	4440	34800	88	1001010011	4510	34500
44	1001100011	4450	34800	90	0111100111	5120	37000
45	1011100110	4660	33000	93	1001010011	4510	34500
46	1001111011	4230	35600	94	1011010001	5010	32700
47	0111010011	5180	36700	95	0001101111	4170	35900
50	1011100110	4660	33000	97	1011100110	4660	33000
54	0011101011	4170	35900	98	1001011110	4300	37200
56	0111100011	5120	37000	99	1001011110	4300	37200

Source: Own work based on Guze (2015)
Note: Chromosomes constitute Pareto front ID: 2, 11, 23, 28, 47, 56, 57, 63, 64, 75, 79, 83, 90, 94

Table B.3
Results for Case 3

ID	Chromosom	MTTF [h]	Cost [USD]	ID	Chromosom	MTTF [h]	Cost [USD]
4	1011100110	4660	33000	52	1011101010	4380	34100
6	1001101010	4240	37500	53	1011100110	4660	33000
7	1011010110	4720	32700	54	1011010110	4720	32700
8	1111010111	5180	36700	57	1011101010	4380	34100
11	1011101010	4380	34100	58	1011100110	4660	33000
12	1011011110	4440	33800	59	1001101010	4240	37500
13	1001011110	4300	37200	60	0011101010	4380	34100
15	1011010101	5010	32700	61	1011101010	4380	34100
18	1011101010	4380	34100	64	1011101010	4380	34100
20	1011010110	4720	32700	65	1011010110	4720	32700
25	1011101010	4380	34100	69	1011011110	4440	33800
26	1001101010	4240	37500	70	1011100110	4660	33000
27	1011010110	4720	32700	72	1011000111	4480	33900
28	1001011010	4300	37200	73	1001011110	4300	37200
29	1011010110	4720	32700	74	1011010010	4720	32700
31	1001011010	4300	37200	76	1011101010	4380	34100
32	1001011110	4300	37200	77	1001011110	4300	37200
33	1011010110	4720	32700	80	1011000111	4480	33900
34	1001011110	4300	37200	82	1011101110	4380	34100
36	1011101010	4380	34100	83	1011010001	5010	32700
43	1011010110	4720	32700	84	1011100110	4660	33000
44	1011100110	4660	33000	86	1011100110	4660	33000
45	1011010110	4720	32700	94	1011000111	4480	33900
47	1011101010	4380	34100	96	1111010111	5180	36700
51	1011000111	4480	33900	99	1001011110	4300	37200

Source: Own work based on (Guze, 2015)
Note: Chromosomes constitute Pareto front ID: 4, 7, 8, 15, 20, 27, 29, 33, 43, 44, 45, 53, 54, 58, 65, 70, 74, 83, 84, 86, 96

Table B.4
Results for Case 4

ID	Chromosom	MTTF [h]	Cost [USD]	ID	Chromosom	MTTF [h]	Cost [USD]
1	0011110111	4620	32000	52	1011101010	4380	34100
5	1011010111	4650	31100	53	0111100111	5120	37000
6	0101101110	3830	35000	55	0101110111	4620	32000
8	1111010111	5180	36700	56	0101110111	4620	32000
9	1111010111	5180	36700	59	1011010111	4650	31100
10	1011010110	4720	32700	61	1101110111	4620	32000
11	0101110111	4620	32000	64	1001110111	4620	32000
12	0101110111	4620	32000	65	1011010111	4650	31100
13	1001101111	4170	35900	67	1011010111	4650	31100
19	1101110111	4620	32000	69	1001101110	4240	37500
22	1111010111	5180	36700	70	1111010111	5180	36700
28	0011110111	4620	32000	71	1111010111	5180	36700
31	1011010111	4650	31100	72	0111110111	4620	32000
32	1001010011	4510	34500	74	0001110111	4620	32000
33	1101110111	4620	32000	75	0101110111	4620	32000
34	0011110111	4620	32000	79	1101110111	4620	32000
35	0101110111	4620	32000	80	1111110111	4620	32000
36	1011101010	4380	34100	81	1001101110	4240	37500
37	0011110111	4620	32000	82	1001101110	4240	37500
43	0011110111	4620	32000	83	0011110111	4620	32000
44	0011110111	4620	32000	86	1001101110	4240	37500
45	1101110111	4620	32000	88	0111110111	4620	32000
48	1011110111	4620	32000	94	0111111111	4230	35600
49	0011101110	4380	34100	95	1101110111	4620	32000
51	1011010111	4650	31100	98	1101110111	4620	32000

Source: Own work based on Guze (2015)
Note: Chromosomes constitute Pareto front ID: 1, 5, 8, 9, 10, 11, 12, 19, 22, 28, 31, 33, 34, 35, 37, 43, 44, 45, 48, 51, 53, 55, 56, 59, 61, 64, 65, 67, 70, 71, 72, 74, 75, 79, 80, 83, 88, 95, 98

Table B.5
Results for Case 5

ID	Chromosom	MTTF [h]	Cost [USD]	ID	Chromosom	MTTF [h]	Cost [USD]
0	0101101010	3830	35000	36	1011101010	4380	34100
1	0011110111	4620	32000	37	1011100010	4660	33000
3	0101101011	4170	35900	38	0101110011	4620	32000
5	1011010111	4650	31100	39	1111011111	4230	35600
6	1011010111	4650	31100	40	0001111011	4230	35600
7	0101101010	3830	35000	41	0111100011	5120	37000
12	1010110111	4690	28600	42	1011010111	4650	31100
13	1111011001	4180	34700	43	1011010101	5010	32700
14	1011100010	4660	33000	45	0011111111	4230	35600
16	1001111111	4230	35600	46	1001111011	4230	35600
17	0101110111	4620	32000	47	1111011111	4230	35600
18	0011110111	4620	32000	48	1011010101	5010	32700
19	1101110111	4620	32000	49	1010110111	4690	28600
20	0101101011	4170	35900	51	0101101010	3830	35000
22	0101101011	4170	35900	52	1011101010	4380	34100
23	1011010111	4650	31100	53	1101110111	4620	32000
27	1111011111	4230	35600	54	0101101010	3830	35000
28	0011110111	4620	32000	55	0111101011	4170	35900
29	1101110111	4620	32000	56	0111101011	4170	35900
30	1001111011	4230	35600	57	0001110111	4620	32000
31	1011010101	5010	32700	61	0101101010	3830	35000
32	0101101011	4170	35900	62	1101110111	4620	32000
33	1001110111	4620	32000	63	0011110111	4620	32000
34	1011010101	5010	32700	65	1011010111	4650	31100
35	1101110111	4620	32000	67	0011111111	4230	35600

Source: Own work based on Guze (2015)
Note: Chromosomes constitute Pareto front ID: 1, 5, 6, 12, 14, 17, 18, 19, 23, 28, 29, 31, 33, 34, 35, 37, 38, 41, 42, 43, 48, 49, 53, 57, 62, 63, 65

Bibliography

Agrawal, A. and R. E. Barlow (1984). "A Survey of Network Reliability and Domination Theory". *Operations Research* 32, 478–492.

Aki, S. and K. Hirano (1996). "Lifetime Distribution and Estimation Problems of Consecutive k-out-of-n : F Systems." *Annals of the Institute of Statistical Mathematics* 48, 185–199.

Appert, M. and L. Chapelon. 2007. "Measuring Urban Road Network Vulnerability Using Graph Theory: The Case of Montpellier's Road Network." *La Mise en Carte des Risques Naturels*, 89 pp.

Barabási, A.-L. 2009. "Scale-Free Networks: A Decade and Beyond." *Science* 325(5939): 412–413.

Barabási, A.-L. and R. Albert. 1999. "Emergence of Scaling in Random Networks." *Science* 286(5439): 509–512.

Barlow, R. E. and F. Proschan. 1975. "Importance of System Components and Fault Tree Events." *Stochastic Processes and Their Applications* 3 (2): 153–173.

Bell, M. G. H. and Y. Iida. 1997. *Transportation Network Analysis*. Chichester, West Sussex: Wiley.

Berdica, K. 2002. "Vulnerability: A Model-Based Case Study of the Road Network in Stockholm." PhD diss., TraVIS for Roads.

Berdica, K. and J. Eliasson. 2004. "Regional Accessibility Analysis from a Vulnerability Perspective." *In Proceedings of the Second International Symposium on Transportation Network Reliability (INSTR)*, edited by A. Nicholson and A. Dantas, 89–94. Christchurch, New Zealand.

Bhattacharya, A. and G. R. Vijayakumar. 2002. "Effect of Edge-Subdivision on Vertex-Domination in a Graph." *Discussiones Mathematicae Graph Theory* 22 (2), 335–347.

Blokus-Roszkowska, A. "Reliability Analysis of the Bulk Cargo Loading System Including Dependent Components." *In Proc. The International Conference of Numerical Analysis and Applied Mathematics – ICNAAM 2015, Symposium/Workshop on Safety of Critical Infrastructures*, edited by 1738A, 440002. Rhodes, Greece, 2015. AIP Conf. Proc. 1738.

Cascetta, E. "Transportation Systems in Transportation Systems Engineering: Theory and Methods." *In Transportation Systems in Transportation Systems Engineering*, edited by E. Cascetta, 1–22. Boston: Springer US, MA, 2001.

Chen, A., H. Yang, H. K. Lo, and W. H. Tang. "Capacity Reliability of a Road Network: An Assessment Methodology and Numerical Results." *Transportation Research Part B: Methodological* 36, no. 3 (2002): 225–52. https://doi.org/10.1016/S0191-2615(00)00048-5

Cormen, T. H., C. E. Leierson, R. L. Rivest, and C. Stein. *Introduction to Algorithms*. 3rd ed. MIT Press, 2009. Section 23.2: The Algorithms of Kruskal and Prim, 631–38. ISBN: 0-262-03384-4.

Crowder, M. A Multivariate Distribution with Weibull Connections. *"Journal of the Royal Statistical Society: Series B (Methodological)* 51, no. 1 (1989): 93–107. https://doi.org/10.1111/j.2517-6161.1989.tb01751.x

Cui, D., Z. Y. Gao, and X. M. Zhao. "Cascades in small-world modular networks with 'CML's method." *Physica B* 17 (2007): 1703–10.

D'Este, G. M. and M. A. P. Taylor. "Network Vulnerability: An Approach to Reliability Analysis at the Level of National Strategic Transport Networks." *In The Network Reliability of Transport. Proceedings of the 1st International Symposium on Transportation Network Reliability (INSTR)*, edited by M. G. H. Bell and Y. Iida, 23–44. Oxford, England: Pergamon, 2003.

Diestel, R. *Graph Theory*. New York. Springer, 2000. ISBN: 0-387-95014-1.

Dixit, A. K. *Optimization in Economic Theory*. Oxford University Press, 1990.

Du, Z. P. and A. Nicholson. "Degradable Transportation Systems: Sensitivity and Reliability Analysis." *Transportation Research Part B: Methodological* 31, no. 3 (1997): 225–37. https://doi.org/10.1016/S0191-2615(96)00023-9.

Einarsson, S. and M. Rausand. "An Approach to Vulnerability Analysis of Complex Industrial Systems." *Risk Analysis* 18, no. 5 (1998): 535–46.

Findeisen, W., J. Szymanowski, and A. Wierzbicki. Teoria i Metody Obliczeniowe Optymalizacji. PWN, 1980.

Fink, J. F., M. S. Jacobson, L. F. Kinch, and J. Roberts. "The Bondage Number of Graph." *Discrete Mathematics* 86, no. 1–3 (1990): 47–57.

Ford, L. R., and D. R. Fulkerson. "Maximal Flow through a Network." *Canadian Journal of Mathematics* 8 (1956): 399–404.

Gross, J. L. and J. Yellen. *Handbook of Graph Theory*. CRC Press, 2004. ISBN: 978-1-58488-090-5.

Guze, S. "Reliability Analysis of Multi-State Ageing Series-Consecutive 'm out of n: F' Systems." *In Proc. European Safety and Reliability Conference – ESREL 2009, vol. 3*, 1629–35. Prague, 2009.

Guze, S. "Application of the Knapsack Problem to Reliability Multi-Criteria Optimization." *JPSRA* 5, no. 1 (2014a): 85–90.

Guze, S. "The Graph Theory Approach to Analyze Critical Infrastructures of Transportation Systems." *JPSRA* 5, no. 2 (2014b): 57–62.

Guze, S. "Graph Theory Approach to Transportation Systems Design and Optimization." *TransNav, The International Journal on Marine Navigation and Safety of Sea Transportation* 8, no. 4 (2014c): 571–78. https://doi.org/10.12716/1001.08.04.12.

Guze, S. "Numerical Application of the SPEA Algorithm to Reliability Multiobjective Optimization." *JPSRA* 6, no. 1 (2015): 101–14.

Guze, S. "An Application of the Selected Graph Theory Domination Concepts to Transportation Networks Modelling." *Zeszyty Naukowe Akademii Morskiej w Szczecinie* 52, no. 124 (2017): 97–102.

Guze, S. A Computer-Aided Approach to Analysis and Optimization of Transportation Networks Based on Graph Theory Methods and Algorithms. *Zeszyty Naukowe Uniwersytetu Morskiego w Gdyni*, 2018.

Guze, S. "Business Availability Indicators of Critical Infrastructures Related to Climate-Weather Change." In Contemporary Complex Systems and Their Dependability. DepCoS-RELCOMEX 2018, edited by W. Zamojski, J. Mazurkiewicz, J. Sugier, T. Walkowiak, and J. Kacprzyk, vol. 761 of Advances in Intelligent Systems and Computing. Cham: Springer, 2019.

Guze, S. Modeling and Optimization of Transport Networks Regarding Their Reliability and Sensitivity. Szczecin: Wydawnictwo Maritime University of Szczecin, 2019. (In Polish)

Guze, S. and K. Kołowrocki. "Safety Modeling of Port, Shipping and Ship Traffic and Port Operation Information Critical Infrastructure Joint Network Related to Its Operation Process." In *Proceedings of the European Safety and Reliability Conference (ESREL) 2017*, 751–59. CRC Press/Balkema; Taylor & Francis Group, 2017.

HackerEarth. "Graph Representation Tutorial." Accessed January 13, 2019.

Harary, F. *Graph Theory*. Reading, MA: Addison-Wesley, 1969.

Hartnell, B. L. "Firefighter! An Application of Domination." *Presented at the 20th Conference on Numerical Mathematics and Computing, University of Manitoba, Winnipeg, Canada, September 1995.*

Haynes, T. W., M. Hedetniemi, and T. Hedetniemi. "Domination and Independence Subdivision Numbers of Graphs." *Discussiones Mathematicae Graph Theory* 20 (2000): 271–80.

Haynes, T. W., S. Hedetniemi, and P. Slater. *Fundamentals of Domination in Graphs*. CRC Press, 1998.

Hollnagel, E., D. D. Woods, and N. Leveson. *Resilience Engineering: Concepts and Precepts*. Aldershot: Ashgate Publishing, 2006.

Holmgren, A. *Vulnerability Analysis of Electrical Power Delivery Networks. Licentiate thesis TRITA-LWR LIC 2020, Department of Land and Water Resources Engineering, KTH*, Stockholm, 2004a.

Holmgren, A. J. *"A Framework for Vulnerability Assessment of Electric Power Systems,"* 31–55. Berlin, Heidelberg: Springer Berlin Heidelberg, 2007.

Holmgren, A. J. and S. Molin. "Using Disturbance Data to Assess Vulnerability of Electric Power Delivery Systems." *Journal of Infrastructure Systems* 12, no. 4 (2006): 243–51.

Holmgren, J. (2004b). Holmgren, J. Efficient Updating Shortest Path Calculations for Traffic Assignment. Master's thesis, LITH-MAI-EX-2004-13, Department of Mathematics, Linköping Institute of Technology, Linköping, 2004b.

Hongwei, M. and Z. Xizhao. "An Evaluation Method for the Connectivity Reliability Based on the Transportation Network of Critical Links." *International Journal of Transportation* 3, no. 2 (2015): 45–52. https://doi.org/10.14257/ijt.2015.3.2.04.

Hurwicz, L. "Optimality Criteria for Decision Making under Ignorance." *Statistics* 1951: 370. Accessed December 1, 2024. https://cowles.yale.edu/sites/default/files/files/pub/cdp/s-0370.pdf.

Iida, Y. "Basic Concepts and Future Directions of Road Network Reliability Analysis." *Journal of Advanced Transportation* 33, no. 2 (1999): 125–34.

Jackson, C. J., H. Heath, A. Kunst, N. Thom, H. Welton, H. N. J. Tuffaha, and E. Wilson. "The Expected Value of Perfect or Partial Perfect Information." *In Value of Information for Healthcare Decision-Making*, edited by A. Heath, N. Kunst, and C. Jackson, 47–91. New York: Chapman and Hall/CRC, 2024.

Jasny, B. R., L. M. Zahn, and E. Marshall. "Connections." *Science* 325, no. 5939 (2009): 405.

Jenelius, E., T. Petersen, and L. Mattsson. "Importance and Exposure in Road Network Vulnerability Analysis." *Transportation Research Part A* 40 (2006): 537–60.

Johnson, D. S. and L. A. McGeoch. "The Traveling Salesman Problem: A Case Study in Local Optimization." *In Local Search in Combinatorial Optimisation*, edited by E. H. L. Aarts and J. K. Lenstra, 215–310. Chichester: John Wiley and Sons, 1997.

Kalashinkov, V. V., S. Dempe, G. A. Perez-Valdes, N. Kalashnykova, and J.-F. Comacho-Vallejo. "Bilevel Programming and Applications, Mathematical Problems in Engineering." *Mathematical Problems in Engineering (2015)*: Article ID 310301, 16 pages. https://doi.org/10.1155/2015/310301.

Khan, M. A. "Transportation Cost Optimization Using Linear Programming." *In International Conference on Mechanical, Industrial and Energy Engineering*, 1–5. Khulna, Bangladesh, December 2014.

Kołowrocki, K. *Reliability of Large Systems*. Amsterdam–Boston–Heidelberg –London–New York–Oxford–Paris–San Diego–San Francisco–Singapore–Sydney–Tokyo: Elsevier, 2004.

Kołowrocki, K. "Safety of Critical Infrastructures." *Journal of Polish Safety and Reliability Association, Summer Safety and Reliability Seminars* 4, no. 1 (2013): 51–74.

Kołowrocki, K. and J. Soszyńska-Budny. *Reliability and Safety of Complex Technical Systems and Processes: Modeling – Identification – Prediction – Optimization. London*; Dordrecht; Heidelberg; New York: Springer, 2011

Kuang, A., Z. Tang, and L. Shan. "Road Network Capacity Reliability Considering Travel Time Reliability: 13th COTA International Conference of Transportation Professionals (CICTP 2013)." *Procedia – Social and Behavioral Sciences* 96 (2013): 1818–27.

Lam, W. H. K. and N. Zhang. "A New Concept of Travel Demand Satisfaction Reliability for Assessing Road Network Performance." *In Proceedings of the Matsuyama Workshop on Transport Network Analysis*, 165–78. Matsuyama, Japan, 2000.

Lawler, E., J. Lenstra, A. Rinnooy Kan, and D. Shmoys. *The Traveling Salesman Problem: A Guided Tour of Combinatorial Optimization*. New York: Wiley, 1985.

Lee, L. "Multivariate Distributions Having Weibull Properties." *Journal of Multivariate Analysis* 9, no. 2 (1979): 267–77. https://doi.org/10.1016/0047-259X(79)90084-8

Leeuwen, V. J. "Graph Algorithms Book." *In Handbook of Theoretical Computer Science*, 525–631. 1990.

Lu, J. and G. K. Bhattacharyya. "Some New Constructions of Bivariate Weibull Models." *Annals of the Institute of Statistical Mathematics* 42, no. 3 (1990): 543–59.

McNamee, P. and J. Celona. Decision Analysis for the Professional. SmartOrg, Inc., 2008.

Mine, H. and H. Kawai. *Mathematics for Reliability Analysis*. Tokyo: Asakura-Shoten, 1982.

Ming-Hua, L., T. Jung-Fa, and Y. Chian-Son. "A Review of Deterministic Optimization Methods in Engineering and Management." *Mathematical Problems in Engineering* (2012).

Navarro, J. and S. Eryilmaz. "Mean Residual Lifetimes of Consecutive-k-out-of-n Systems." *Journal of Applied Probability* 44, no. 1 (2007): 82–98. https://doi.org/10.1239/jap/1175267165.

Nelder, J. A. and R. Mead. "A Simplex Method for Function Minimization." *The Computer Journal* 7 (1965): 308–13.

Nešetřil, J., E. Milková, and H. Nešetřilová. "Otakar Borůvka on Minimum Spanning Tree Problem: Translation of Both the 1926 Papers, Comments, History." *Discrete Mathematics* 233 (2001): 3–36.

Newell, G. F. *Traffic Flow on Transportation Networks*. Monograph 5, MIT Press Series in Transportation Studies. Cambridge, MA: MIT Press, 1980.

Nowakowski, T. Niezawodność Systemów Logistycznych. Wrocław: Oficyna Wydawnicza Politechniki Wrocławskiej, 2011.

Papadimitriou, C. H. "Games Against Nature." *Journal of Computer and System Sciences* 31, no. 2 (1985): 288–301.

Patriarca, R., J. Bergström, G. Di Gravio, and F. Costantino. "Resilience Engineering: Current Status of the Research and Future Challenges." *Safety Science* 102 (2018): 79–100. https://doi.org/10.1016/j.ssci.2017.10.005

Pazek, K. and E. Rozman. "Decision Making under Conditions of Uncertainty in Agriculture: A Case Study of Oil Crops." *Poljoprivreda* 15, no. 1 (2009): 45–50

Proag, V. 2014. "The Concept of Vulnerability and Resilience." *Procedia Economics and Finance* 18: 369–76. https://doi.org/10.1016/S2212-5671(14)00952-6

Raiffa, H. 1968. Decision Analysis: Introductory Lectures on Choices under Uncertainty. New York: Random House.

Rausand, M. and A. Høyland. 2004. *System Reliability Theory: Models, Statistical Methods and Applications*. 2nd ed. New York: Wiley Series in Probability and Statistics.

Ray, D. 2005. *Optimization for Engineering Design: Algorithms and Examples*. India: Prentice-Hall.

Reggiani, A., P. Nijkamp, and D. Lanzi. 2015. "Transport Resilience and Vulnerability: The Role of Connectivity." *Transportation Research Part A: Policy and Practice* 81: 4–15.

Rodrigue, J. P., C. Comtois, and B. Slack. 2017. *The Geography of Transport Systems*. 4th ed. New York: Routledge, Taylor & Francis Group.

Rosenthal, A. 1977. "Computing the Reliability of Complex Networks." *SIAM Journal on Applied Mathematics* 32 (2): 384–93.

Roy, D. and S. P. Mukherjee. "Characterizations of Some Bivariate Lifed-istributions." *Journal of Multivariate Analysis* 28, no. 1 (1989): 1–8. https://doi.org/10.1016/0047-259X(89)90095-X.

Salas, S., E. Hille, and G. Etgen. *Calculus*. 10th ed. Hoboken, NJ: John Wiley & Sons, 2007.

Sarewitz, D., R. Pielke Jr., and M. Keykhah. "Vulnerability and Risk: Some Thoughts from a Political and Policy Perspective." *Risk Analysis* 23, no. 4 (2003): 805–10.

Šarūnienė, I., L. Martišauskas (from "Martišauskas"), R. Krikštolaitis (from "Krikštolaitis"), J. Augutis, and R. Setola. 2024. "Risk Assessment of Critical Infrastructures: A Methodology Based on Criticality of Infrastructure Elements." *Reliability Engineering and System Safety* 243: 109797.

Savage, L. J. "The Theory of Statistical Decision." *Journal of the American Statistical Association* 46, no. 253 (1951): 55–67.

Shariat Mohaymany, A. and M. Babaei. 2013. Optimal resource allocation in urban transportation networks considering capacity reliability and connectivity reliability: a multi-objective approach. *International Journal of Civil Engineering 11*(1), 33–42. Transaction A: Civil Engineering.

Simon, H. A. "Theories of Decision-Making in Economics and Behavioral Science." *In American Economic Review*, 1959.

Solé, R. V. and S. Valverde. "Information Theory of Complex Networks: On Evolution and Architectural Constraints, Complex Networks." *In Springer*, 189–207. Heidelberg: Berlin, 2004.

Strzelczyk, A. and S. Guze. "The Game Theory Decision Models for Transport Systems Selection Based on the SWOT Analysis on the Example of Heavy and Oversized Goods Transport." *Scientific Journals of the Maritime University of Szczecin* 81, no. 153 (2025): 23–34.

Taylor, M. A. P. and G. M. D'Este. "Critical Infrastructure and Transport Network Vulnerability: Developing a Method for Diagnosis and Assessment." *In Proceedings of the Second International Symposium on Transportation Network Reliability (INSTR)*, edited by A. Nicholson and A. Dantas, 96–102. Christchurch, New Zealand, 2004.

Tinós, R. "Analysis of the Dynamic Traveling Salesman Problem with Weight Changes." In 2015 Latin America Congress on Computational Intelligence (LA-CCI), 1–6. *Curitiba*, 2015. https://doi.org/10.1109/LA-CCI.2015.7435936.

Topp, J. "Graph with Unique Minimum Edge Dominating Sets and Graphs with Unique Maximum Independent Sets of Vertices." *Discrete Mathematics* 121 (1993): 199–210. North-Holland.

Tulsian, P. C. and V. Pandey. *Quantitative Techniques: Theory and Problems*. India: Pearson, 2016.

Velammal, S. *Studies in Graph Theory: Covering, Independence, Domination and Related Topics*. PhD diss., Manonmaniam Sundaranar University, Tirunelveli, 1997.

Venter, G. *Review of Optimization Techniques. In Encyclopedia of Aerospace Engineering*. USA: John Wiley & Sons, 2010.

Wald, A. "Statistical Decision Functions." *The Annals of Mathematical Statistics* (1949): 165–205.

Walikar, H. B., B. D. Acharya, and E. Sampathkumar. "Recent Developments in the Theory of Domination in Graphs." *Allahabad*, 1 (1979).

Watts, D. J. and S. H. Strogatz. "Collective Dynamics of 'Small-World' Networks." *Nature* 393, no. 6684 (1998): 440–42.

Willie, R. R. Computer-Aided Fault Tree Analysis. ORC 78-14, Operations Research Center, University of California, Berkeley, 1978.

Wilson, R. J. Introduction to Graph Theory. 1998.

Zambrano-Martinez, J. L., C. T. Calafate, D. Soler, J. C. Cano, and P. Manzoni. "Modeling and Characterization of Traffic Flows in Urban Environments." *Sensors (Basel)* 18, no. 7 (2020). https://doi.org/10.3390/s18072020

Zhan, F. B. and C. E. Noon. "Shortest Path Algorithms: An Evaluation Using Real Road Networks." *Transportation Science* 32, no. 1 (1998): 65–73.

Zitzler, E. and L. Thiele. "Multiobjective Evolutionary Algorithms: A Comparative Case Study and the Strength Pareto Approach." *IEEE Transactions on Evolutionary Computation* 3, no. 4 (1999): 257–71.

Index